Horizons in World Physics
Volume 248

SPACETIME PHYSICS RESEARCH TRENDS

Horizons in World Physics

Volume 248 **Spacetime Physics Research Trends**
Albert Reimer

Volume 247 **New Developments in Quantum Cosmology Research**
Albert Reimer

Volume 246 **Quantum Cosmology Research Trends**
Albert Reimer

Volume 245 **Horizons in World Physics**
Tori V. Lynch

Volume 244 **Horizons in World Physics**
Albert Reimer

Volume 243 **Theoretical Physics 2002, Part 2**
Henk F. Arnoldus and Thomas F. George

Volume 242 **Models and Methods of High-T_c Superconductivity: Some Frontal Aspects, Volume 2**
Edited by J. K. Srivastava and S. M. Rao

Volume 241 **Models and Methods of High-T_c Superconductivity: Some Frontal Aspects, Volume 1**
Edited by J. K. Srivastava and S. M. Rao

Volume 240 **Horizons in World Physics**
Albert Reimer

Volume 239 **Theoretical Physics 2002, Part 1**
Henk F. Arnoldus and Thomas F. George

Volume 238 **Theoretical Physics 2001**
Henk F. Arnoldus and Thomas F. George

Volume 237 **Generalized functions in mathematical physics: Main ideas and concepts**
A.S. Demidov

Volume 236 **Edge Excitations of Low-Dimensional Charged Systems**
Oleg Kirichek

Volume 235 **On the Structure of the Physical Vacuum and a New Interaction in Nature (Theory, Experiment, Applications)**
Yu. A. Baurov

Volume 234 **Electron Scattering on Complex Atoms (Ions)**
V. Lengyel, O. Zatsarinny and E. Remeta

Volume 233 **An Introduction to Hot Laser Plasma Physics**
A.A. Andreev, A.A. Mak and N.A. Solovyev

Volume 232 **Electron-Atom Scattering: An Introduction**
Maurizio Dapor

Volume 231 **Theory of Structure Transformations in Non-equilibrium Condensed Matter**
Alexander L. Olemskoi

Volume 230 **Topics in Cosmic-Ray Astrophysics**
M. A. DuVernois (Ed.)

Volume 229 **Interactions of Ions in Condensed Matter**
A. Galdikas and L. Pranevièius

Volume 228 **Optical Vortices**
Edited by M. Vasnetsov and K. Staliunas

Volume 227 **Frontiers in Field Theory, Quantum Gravity and Strings**
R.K. Kaul, J. Maharana, S. Mukhi and S. K. Rama

Volume 226 **Quantum Groups, Noncommutative Geometry and Fundamental Physical Interactions**
D. Kastler, M. Russo and T. Schücker

Volume 225 **Aspects of Gravitational Interactions**
S.K. Srivastava and K.P. Sinha

Volume 224 **Dynamics of Transition Metals and Alloys**
S. Prakash

Volume 223 **Fields and Transients in Superhigh Pulse Current Devices**
G.A. Shneerson

Volume 222 **Gaussian Beams and Optical Resonators**
A. N. Oraevsky

Volume 221 **Laser Cathode-Ray Tubes**
Edited by Yu. M. Popov

Volume 220 **Quantum Electrodynamics with Unstable Vacuum**
Edited by V.L. Ginzburg

Volume 219 **The Muon Method in Science**
V.P. Smilga and Yu.M. Belousov

Volume 218 **Radiative and Collisional Characteristics of Ions in Hot Plasmas**
Edited by I.I. Sobel'man

Volume 217	**Nonlinear and Quantum Optical Phenomena in Nonequilibrium Media** Edited by V.A. Shcheglov
Volume 216	**Semiconductors and Insulators: Optical and Spectroscopic ...** Edited by Yu.I. Koptev
Volume 215	**Relativistic Theory of Gravity** A.A. Logunov
Volume 214	**Research in Quantum Field Theory** Edited by V.I. Man'ko
Volume 213	**Nonlinear Theory of Strong Electromagnetic Wave-plasma...** Edited by O.N. Krokhin
Volume 212	**Dynamics of Elementary Atomic-Molecular...** Edited by O. S. Scheglov
Volume 211	**Optics and Lasers** Edited by G.G. Petrash
Volume 210	**Theory and Technology of High Temperature Superconductors** Edited by Yu.I. Koptev
Volume 209	**Theory of Interaction of Multilevel Systems...** Edited by V.I. Man'ko
Volume 208	**Physical Properties of High Temperature Superconductors** Edited by Yu.I. Koptev
Volume 207	**Generalized Mayer Series in Classical Statistical Mechanics** Igor J. Ivanchik
Volume 206	**Mathematical Physics, Applied Mathematics and Informatics** Edited by Yu.I. Koptev
Volume 205	**Squeezed and Correlated States of Quantum Systems** Edited by M.A. Markov
Volume 204	**Sakharov Memorial Lectures in Physics** Edited by L.V. Keldysh and V.Ya. Fainberg
Volume 203	**Fiber Optics: Research and Development** Edited by Ye.M. Dianov
Volume 202	**Gas Dynamics** Edited by Yu.I. Koptev
Volume 201	**Generation of Nonlinear Waves and Quasistationary ...** Edited by L.M. Kovrizhnykh
Volume 200	**Laser Thermonuclear Targets and Superdurable Microballoons** Edited by A.I. Isakov
Volume 199	**Spatiotemporal Characteristics of Laser Emission** M.V. Pyatakhin and A.F. Suchkov
Volume 198	**Theory of Nonstationary Quantum Oscillators** Edited by M.A. Markov
Volume 197	**Reliability Problems of Semiconductor Lasers** P.G. Eliseev
Volume 196	**Pulsars** Edited by A.D. Kuzmin
Volume 195	**Physical Characteristics and Critical Temperature of High Temperature Superconductors** Edited by M.M. Sushchinskiy
Volume 194	**Biophysical Approach to Complex Biological Phenomena** Edited by E. Volkov
Volume 193	**Research on Chemical Lasers** Edited by A.N. Orayevskiy
Volume 192	**Atomic and Ionic Spectra and Elementary Processes in Plasma** Edited by I.I. Sobelman
Volume 191	**Polarization of the Vacuum and a Quantum Relativistic Gas ...** A.Ye. Shabad
Volume 190 (Supp. Vol.)	**Superconductors with A15 Lattice and Bridge Contacts Based ...** Edited by M. M. Sushchinskiy
Volume 190	**Metal-Optics and Superconductivity** Edited by A.I. Golovashkin
Volume 189	**Clusters of Galaxies and Extragalactic Radio Sources** Edited by A.D. Kuz'min
Volume 188	**Integral Systems, Solid State Physics and Theory of Phase Transitions. Pt. 2. Symm. and Alg. Structures in Physics** Edited by V.V. Dodonov and V.I. Man'ko
Volume 187	**Quantum Field Theory, Quantum Mechanics and Quantum Optics: Pt. 1. Symm. and Alg. Structures in Physics** Edited by V.V. Dodonov and V.I. Man'ko

Volume 186	**Photoproduction of Pions on Nucleons and Nuclei** Edited by A.A. Komar
Volume 185	**Injection Lasers in Optical Communication and Information Processing Systems** Edited by Yu. M. Popov
Volume 184	**Electron Processes in MIS-Structure Memories** Edited by A.F. Plotnikov
Volume 183	**Invariants and the Evolution of Nonstationary Quantum Systems** Edited by M.A. Markov
Volume 182	**Luminescence of Wideband Semiconductors** Edited by M.D. Galanin
Volume 181	**Pulsed Metal and Metal Halogenide Vapor Lasers** Edited by G.G. Petrash
Volume 180	**Inelastic Light Scattering in Crystals** Edited by M.M. Sushchinskiy
Volume 179 (Supp. Vol. 2)	**Electron-Excited Molecules in Nonequilibrium Plasma** Edited by N.N. Sobolev
Volume 179 (Supp. Vol. 1)	**Interaction of Ultrashort Pulses with Matter** Edited by M.D. Galanin
Volume 179	**X-Ray Plasma Spectroscopy and the Properties of Multiply-Charged Ions** Edited by I.I. Sobel'man
Volume 178	**The Delfin Laser Thermonuclear Installation: Operational Complex ...** Edited by G.V. Sklizkov
Volume 177	**Stoichiometry in Crystal Compounds and Its Influence on Their Physical Properties** Edited by Yu. M. Popov
Volume 176 (Supp. Vol)	**Quantization, Gravitation and Group Methods in Physics** Edited by A.A. Komar
Volume 176	**Classical and Quantum Effects in Electrodynamics** Edited by A.A. Komar
Volume 175 (Supp. Vol.)	**Luminescence and Anisotropy of Zinc Sulfide Crystals** Edited by M.D. Galanin
Volume 175	**Luminescence Centers of Rare Earth Ions in Crystal Phosphors** Edited by M.D. Galanin
Volume 174 (Supp. Vol. 2)	**Electron Liquid Theory of Normal Metals** Edited by V.P. Silin
Volume 174 (Supp. Vol. 1)	**Nonequilibrium Superconductivity** Edited by V.L. Ginzburg
Volume 174	**Thermodynamics and Electrodynamics of Superconductors** Edited by V.L. Ginzburg
Volume 173	**Quantum Mechanics and Statistical Methods** Edited by M.M. Sushchinskiy
Volume 172	**Phase Conjugation of Laser Emission** Edited by N.G. Basov
Volume 171	**Research on Laser Theory** Edited by A.N. Orayevskiy
Volume 170	**The Theory of Target Compression by Longwave Laser Emission** Edited by G.V. Sklizkov
Volume 169	**The Physical Effects in the Gravitational Field Black Holes** Edited by M.A. Markov
Volume 168	**Issues in Intense-Field Quantum Electrodynamics** Edited by V.L. Ginzburg
Volume 167	**Group Theory, Gravitation and Elementary Particle Physics** Edited by A.A. Komar
Volume 166	**The Nonlinear Optics of Semiconductor Lasers** Edited by N.G. Basov
Volume 165	**Solitons and Instantons, Operator Quantization** Edited by V.L. Ginzburg

Horizons in World Physics
Volume 248

SPACETIME PHYSICS RESEARCH TRENDS

Albert Reimer
Editor

Nova Science Publishers, Inc
New York

For permission to use material from this book please contact us:
Telephone 631-231-7269; Fax 631-231-8175
Web Site: http://www.novapublishers.com

ISBN 1-59454-322-4

Published by Nova Science Publishers, Inc. ✦ New York

CONTENTS

PREFACE

In the theory of gravitational field the problems connected with the energy conservation exist in a literal sense since the time of its creation when Einstein set up the problem of including gravity into the framework of the Faraday concept of field. Thorough and deep analysis of the problem of gravity field given by him in the works, enables to formulate the key principles of gravity physics (General Relativity). However, till now in the framework of these principles there is no adequate solution to the energy conservation problem . This topic is still the subject of active research and still not completely understood. In particular, the question of whether one can ascribe an energy density to the gravitational field as one can to fields in special relativity is still open. In the first chapter, *Field Theory and the Essence of Time*, Ivanhoe Pestov gives a simple solution of the problem in question, which is based on the connection between the time and energy and necessarily follows from the first principles of General Relativity if one puts them into definite logical sequence. The energy conservation means that the rate of change with time of total energy density of gravity field and all other fields is equal to zero and hence energy density represents the first integral of the system.

The absence of time in the Wheeler-DeWitt (constraint) equation is the main point leading to the generally accepted conclusion that in the quantum cosmology there is no possibility to describe the evolution of the universe proceeding in the cosmic time (the time usually used in classical cosmology). E. I. Guendelman and A. B. Kaganovich, in *Cosmic Time Paradox in Quantum Cosmology*, show that there exist anisotropic quantum cosmology models where in spite of the constraint, the averaging of some of the Heisenberg equations yields *explicit cosmic time dependence of the expectation values of certain dynamical variables*. The reason is that in the context of the minisuperspace approach to quantum cosmology, anisotropic solutions of the Wheeler-DeWitt equation have a feature similar to that of a free nonrelativistic particle solution. The latter implies that these states can be normalized only in a continuous way and a careful definition of the averaging procedure is needed. This "technical detail" leads to the mentioned above appearance of explicit cosmic time dependence of the expectation values of certain dynamical variables, - the result which can be treated at least, as a particular resolution of the conceptual problem of time in quantum gravity. This idea is realized explicitly in two higher dimensional models: in the first model, the sources of gravity are a negative cosmological constant and dust; in the second model - a negative cosmological constant and coherent dilaton excitations. The authors find the new quantum effect which consists in "quantum inflationary phase" for some dimensions and simultaneous "quantum deflationary" contraction for the remaining dimensions. The expectation value of the "volume" of the universe remains constant during this quantum "inflation-deflation" process. They briefly mention the works of other authors (A. Kheyfets,

W.A. Miller, A.P. Gentle and N.D. George; M. Weinstein and R. Akhoury) which also show a possibility of an explicit cosmic time dependence of the expectation values of some of the cosmological variables.

In Chapter 3 *Cover Schemes, Frame-Valued Sets and Their Potential Uses in Spacetime Physics*, by John L. Bell, the concept of a covering is presented and developed. The relationship between cover schemes, frames (complete Heyting algebras), Kripke models, and frame-valued set theory is discussed. Finally cover schemes and frame-valued set theory are applied in the context of Markopoulou's account of discrete spacetime as sets "evolving" over a causal set. We observe that Markopoulou's proposal may be effectively realized by working within an appropriate frame-valued model of set theory. We go on to show that, within this framework, cover schemes may be used to force certain conditions to prevail in the associated models: for example, rendering the universe timeless, obliterating a given event or forcing it to become the universe's "beginning".

Since the nineteenth century, the problem of the arrow of time has been traditionally analyzed in terms of entropy by relating the direction past-to-future to the direction of the entropy increase in isolated systems. In *A Global and Non-Entropic Approach to the Problem of the Arrow of Time,* Mario Castagnino and Olimpia Lombardi reject this traditional perspective and argue for a global and non-entropic approach to the problem, according to which the arrow of time can be defined in terms of the geometrical properties of spacetime. In particular, they show how the global non-entropic arrow can be transferred to the local level, where it takes the form of a non-spacelike local energy flow that provides the criterion for breaking the symmetry resulting from time-reversal invariant local laws.

In Chapter 5, *Light in Metric Space-Time and its Deflection by the Screw Dislocation*, Miroslav Pardy explains the geometrical meaning of the metric of space-time. He also discusses the problem of the dimensionality of space. He argues, by means of some experimental facts, such as the molecule DNA, the galaxy formation, QED, Planck's law and so on that the external space is 3-dimensional and its dimensionality cannot be proved a priori, in accordance with the Russell philosophy of mathematics. Then, he derives the light deflection caused by the screw dislocation in space-time. The derivation is based on the definition of the screw dislocation and on the idea that space-time is a medium which can be deformed in such a way that the deformation of space-time is equivalent to the existence of metric which is equivalent to gravity. The idea that space-time is a medium is tested by the gravitational time dilation in the nonrelativistic limit. The existence of the screw dislocation in the cosmology is hypothetically confirmed by observation of light or gamma cosmical bursts which can be interpreted as the annihilation of the giant (screw) dislocations with the anti-dislocations where antidislocation is introduced in accordance with particle and string physics, where to every particle exists antiparticle and to every string exists antistring. The origin of gravitational bursts are analogical to the optical ones. Hubble telescope is able to detect only the optical bursts. The gravitational bursts then can be detected by LIGO, VIRGO, GEO, TAMA and so on. He argues that cosmological constant plays substantial role in the formation of the screw dislocations. The existence of dislocations and other topological defects in cosmology modify the distribution of the microwaves in cosmical background. The Misner-Wheeler geometrodynamics is here considered as the alternative form of the topological defect origin of the elementary particles.

In Chapter 6, *Motion of Extended Body on Curved Spacetime*, Akira Ohashi developed a *multipole particle* approach to describe the motion of extended bodies on a curved spacetime.

The concept of the multipole particle is to replace extended bodies with point-like particles and to describe motion of the objects by the particles. The particle can have arbitrary multipole structures and its multipole moments are defined by the gravitational field created by the particle. We construct the energy momentum tensor for such a particle. Once the energy momentum tensor is constructed, we can derive equations of motion for the particle. As some example, we will derive relativistic equations of motion for spinning particle, and discuss the practical applications to the post-Newtonian approximation of Einstein's theory of gravitation.

A. Reimer
November 2005

In: Horizons in World Physics, Volume 248
Editor: A. Reimer, pp. 1-29
ISBN 1-59454-322-4

Chapter 1

FIELD THEORY AND THE ESSENCE OF TIME

Ivanhoe Pestov
Bogoliubov Laboratory of Theoretical Physics,
Joint Institute for Nuclear Research, 141980 Dubna, Russia

ABSTRACT

In the framework of the field theory, it is shown that a time (viewed as a scalar temporal field) is an internal property of the physical system, which defines its causal structure and evolution. A concept of internal time predicts the existence of matter outside the time and allows to solve the energy problem in General Relativity. The law of energy conservation means that the total energy density of the full system of interacting fields (including gravitational field) does not vary with time, thus being the first integral of the system. It is demonstrated that direct introduction of the temporal field permits to derive the physical laws of the electromagnetic field (the general covariant four dimensional Maxwell equations for the electric and magnetic fields) from the geometrical equations of this field. It means that the fundamental physical laws are in full correspondence with the essence of time. On this ground, the concept of the momentum of the gravitational field is introduced and from the geometrical laws of the gravitational field the physical evolution equations of this field are derived. Two characteristic solutions of these equations are obtained (including the Schwarzschild solution). Other one describes the emergence and evolution of the clot of pure gravitational energy. The general covariant definition of the time reversal is given and the modification of the Maxwell equations is considered that is not invariant with respect to the time reversal.

INTRODUCTION

In the theory of gravitational field the problems connected with the energy conservation exist in a literal sense since the time of its creation when Einstein set up the problem of including gravity into the framework of the Faraday concept of field. Thorough and deep analysis of the problem of gravity field given by him in the works [1], enables to formulate the key principles of gravity physics (General Relativity). However, till now in the framework of these principles there is no adequate solution to the energy conservation problem [2,3,4,5].

This topic is still the subject of active research and still not completely understood. In particular, the question of whether one can ascribe an energy density to the gravitational field as one can to fields in special relativity is still open. In the presented paper we give a simple solution of the problem in question, which is based on the connection between the time and energy and necessarily follows from the first principles of General Relativity if one puts them into definite logical sequence. The energy conservation means that the rate of change with time of total energy density of gravity field and all other fields is equal to zero and hence energy density represents the first integral of the system.

New understanding of time presented here may have implications for the problem of time in quantum gravity. In fact, a major conceptual problem in this field is the notion of time and how it should be treated. The importance of this issue was recognized at the beginning of the history of quantum gravity [6], but the problem is still unresolved and has drawn recently an increasing attention (see, for example [7,8,9] and further references therein).

The paper is organized as follows. In section 2 we formulate the fundamentals of general covariant theory of time with the guiding idea that the manifold is the main notion in physics and that time itself is a scalar field on the manifold which defines the evolution of the full system of fields being one of them. The system of interacting fields is considered to be full if the gravitational field is included in it.

All known physical laws of nature have the following form: the rate of change with time of certain quantity equals to the results of action of some operator on this quantity. So, a general covariant definition of rate of change with time of any field is one of the main results of the theory of time presented here. It is also a starting point and relevant condition for the consideration of the concept of evolution and of the problem how to write the field equations in the general covariant evolution form.

In section 3 on the ground of the temporal field the Einstein gravitational potential is established. This gives the possibility to derive the geometrical laws of the gravitational field, general covariant law of energy conservation and produce the general covariant expression for the energy density of the gravity field.

To demonstrate a concrete application of general theory, in section 4 it is shown how evolution equations for the vectors of electric and magnetic fields in the four dimensional general covariant form can be derived from the geometrical equations for the bivector of electromagnetic field. Through this it is shown that the original Maxwell equations (which as a matter of fact express the fundamental physical law) are in full correspondence with fundamentals of the theory of time. In section 5 the notion of the momentum of the gravitational field is introduced and dynamical equations of this field is deduced from the geometrical laws established in section 3. In section 6 the exact solutions of these equations are considered. One of them describes the gravitational field of pointlike gravitational charge and other one represents the field of the homogeneous and isotropic distribution of the gravitational energy. Some concluding remarks are given in section 7.

Central Role of Time in Gravity Theory

In accordance with the fundamental principles of geometry and gravity physics, in this section the fundamentals of the theory of time are formulated.

According to Einstein, in presence of gravitational field all the systems of coordinates are on equal footing and in general, coordinates have neither physical nor geometrical meaning. Thus, in the gravity theory the coordinates play the same role as the Gauss coordinates in his internal geometry of surfaces from which all buildings of the modern geometry are grown. Hence, one needs to construct the internal theory of physical fields analogous to the Gauss internal geometry of surfaces. In other words, the problem is how the modern differential geometry transforms into the physical geometry.

The fields characterize the events and fill in the geometrical space. In view of what has been said above, this space is a four dimensional smooth manifold because this structure does not distinguish intrinsically between different coordinate systems (the principle of general covariance is naturally included into this notion). Hence it follows that the notion of smooth manifold is the primary issue not only in differential geometry but also in the theoretical physics (dealing with gravitational phenomena. This means that all other definitions, notions and laws should be introduced into the theory through the notion of smooth manifold. Indeed if some notion, definition or law is in agreement with the structure of smooth manifold, then they are general covariant, i.e., do not depend on the choice of coordinate system.

Definition of manifold is considered to be known and we only notice that all information on this topic can be found for example in reference [4,5]. For our purposes it is enough to keep in mind that all smooth manifolds can be realized as surfaces in the Euclidean space. In what follows we shall consider only four dimensional manifolds. That is evident from the physical point of view. However there is also a deep purely mathematical reason for this choice. Smooth manifold consists of topological manifold and differential structure defined on it. It is known [10], that a topological manifold always admits differential structure if and only if its dimension is not larger than four.

It is clear that, in general, manifold should be in some relation with its material content i.e., the fields. As it comes out, the relation between them can be viewed as a form and content. In view of this it is very important to know how material content designs manifold. If we take the point of view that manifold *a priori* is arena for the physical events, then it is natural to select simplest manifold, for example, manifold of special theory of relativity. However, Einstein argued (in the papers mentioned above) that nature of gravity field is not compatible with an *a priori* defined manifold.

It can be shown that manifold as a surface in the Euclidean space is designed by the covariant symmetrical tensor field g_{ij} on the manifold, for which adjoined quadratic differential form (Riemann metric)

$$ds^2 = g_{ij} du^i du^j \qquad (1)$$

is positive definite. Thus, covariant symmetrical tensor field $g_{ij}(u)$ is the necessary element of any general covariant intrinsically self-consistent physical theory. Here we only give the defining system of differential equations

$$g_{ij}(u^1, u^2, u^3, u^4) = \delta_{ab} \frac{\partial F^a}{\partial u^i} \frac{\partial F^b}{\partial u^j}, \quad a, b = 1, \cdots, 4 + k, k \geq 0$$

avoiding any detailed consideration. If the functions $g_{ij}(u^1,u^2,u^3,u^4)$ are known in local system of coordinates u^1,u^2,u^3,u^4, then solving this system of equations we obtain the functions $F^a(u^1,u^2,u^3,u^4)$ and hence the region of manifold, defined by the equations

$$x^a = F^a(u^1,u^2,u^3,u^4),$$

where x^a are the Cartesian coordinates of embedding Euclidean space.

An important conclusion that follows from this consideration is that there is one and only one way for the other fields to design manifold which can be explained as follows. Let u^1,u^2,u^3,u^4 be a local system of coordinates in the vicinity of some point of an abstract manifold. Let us determine in such a vicinity the system of differential equations that connect basis field $g_{ij}(u)$ with other ones. Solving this system of equations we find field $g_{ij}(u)$ and by doing so, we design a local manifold of physical system in question. In what follows, a smooth manifold that corresponds to a physical system will be called a physical manifold.

For further consideration of the principles of internal field theory we simply note that there is a fundamental difference between physics and geometry. In geometry there is no motion that is tightly connected with the concept of time. Thus, to be logical, we need to introduce time into the theory using its first principles. Since, in general, coordinates have no physical sense, time can be presented as a set of functions of four independent variables (or in more strict manner as a geometrical object on the manifold). It is quite obvious from the logical point of view.

We put forward the idea that the time is a scalar field on the manifold. By this we get a simple answer to the question with long standing history "What is time ?" It should be emphasized, that temporal field (together with other fields) designs manifold as it was explained above but it has also another functions which will be considered below.

Temporal field with respect to the coordinate system u^1,u^2,u^3,u^4 in the region U of smooth four dimensional manifold M is denoted as

$$f(u) = f(u^1,u^2,u^3,u^4).$$

If the temporal field is known, then to any two points p and q of manifold one can put in correspondence an interval of time

$$t_{pq} = f(q) - f(p) = \int_p^q \partial_i f du^i. \tag{2}$$

Unlike time, space is not an independent entity. Instead of space we shall consider space cross-sections of the manifold $f^{-1}(t)$, which are defined by the temporal field. For the real number t, space cross-section is defined by the equation

$$f(u^1,u^2,u^3,u^4)=t. \tag{3}$$

One can call the number t "the height" of the space cross-section of manifold. If a point p belongs to the space cross-section $f^{-1}(t_1)$, and a point q to the space cross-section $f^{-1}(t_2)$, then the time interval (2) is equal to the difference of the heights $t_{pq}=t_2-t_1$. It is clear that $t_{pq}=0$ if p and q belong to the same space cross-section. Thus, to the every segment of curve one can put in correspondence a time interval.

Given the general covariant definition of time, one should show that it is constructive in all respects. First of all we consider how the temporal field defines the form of physical laws. It is known that the general form of physical laws is very simple and is based on the following recipe: the rate of change with time of a certain quantity is equal to the result of action of some operator on this quantity. To be concrete, let us consider Maxwell equations. We know that the rate of change with time of electrical and magnetic fields enter the dynamical equations of electromagnetic field. Thus, we need to give a general covariant definition of the rate of change with time of any field and in particular this definition should be applicable for the case of electromagnetic field.

This problem has fundamental meaning, since it is impossible to speak about physics when one has no mathematically rigorous definition of evolution. It is clear that correct general covariant definition should be conjugated with simple condition: if the rate of change with time of some field is equal to zero in one coordinate system, then in any other coordinate system the result will be the same.

On the manifold there is only one general covariant operation that can be considered as a basis for the definition of the rate of change with time of any field quantity. This general covariant operation is called derivative in given direction and is defined by the vector field on the manifold and the structure of the manifold itself. Thus, the problem is to connect the temporal field $f(u^1,u^2,u^3,u^4)$ with some vector field $t^i(u^1,u^2,u^3,u^4)$. Since a temporal field is a scalar one, the partial derivatives define covector field $t_i=\partial_i f$. Now, the following definition becomes self-evident: the gradient of temporal field (or the stream of time) is the vector field of the type

$$t^i=(\nabla f)^i=g^{ij}\frac{\partial f}{\partial u^j}=g^{ij}\partial_j f=g^{ij}t_j, \tag{4}$$

where g^{ij} are the contravariant components of the Riemann metric (1). The gradient of the field of time defines the direction of the most rapid increase (decrease) of the field of time. We define now the rate of change with time of some quantity as the derivative in the direction of the gradient of the field of time and denote this operation by the symbol D_t.

Let us find the expression for the rate of change with time of the temporal field itself. We have,

$$D_t f=t^i\partial_i f=g^{ij}\partial_i f\partial_j f.$$

Since $D_t f$ is a general covariant generalization of the evident relation $\frac{d}{dt}t = 1$, the temporal field should obey the fundamental equation

$$(\nabla f)^2 = g^{ij} \frac{\partial f}{\partial u^j} \frac{\partial f}{\partial u^j} = 1. \tag{5}$$

Equation (5) means that the rate of change with time of the temporal field is a constant quantity, the most important constant of the theory. From the geometrical point of view the equation (5) simply shows that the gradient of the temporal field is unit vector field on the manifold with respect to the scalar product that is defined as usual by the metric (1), $(V, W) = g_{ij} V^i W^j$.

It should be noted that the equation $(\nabla f)^2 = 1$ is the main equation of the geometrical optics. In view of this one can consider equation (5) as the equation of 4-optics. This analogy can be useful for consideration of some special problems in the theory of time.

The rate of change with time of the symmetrical tensor field is given by the expression

$$D_t g_{ij} = t^k \frac{\partial g_{ij}}{\partial u^k} + g_{kj} \frac{\partial t^k}{\partial u^i} + g_{ik} \frac{\partial t^k}{\partial u^j}. \tag{6}$$

Similar formulas can be presented for any other geometrical quantities. In mathematical literature the derivative with respect to the given direction is usually called the Lie derivative. Thus, one can say that the rate of change with time of any field is the Lie derivative with respect to the direction of the stream of time.

Consider the notion of time reversal and the invariance with respect to this symmetry that is very important for what follows. It is almost evident that in general covariant form the time reversal invariance means that theory is invariant with respect to the transformations

$$t^i \to -t^i. \tag{7}$$

It is clear that theory will be time reversal invariant if the gradient of temporal fields will appear in all formulae only as an even number of times, like $t^i t^j$.

Within the scope of special relativity and quantum mechanics, time and energy are tightly connected. It is natural to suppose that in gravity physics the link between time and energy even more deep and energy conservation follows from the invariance of the Lagrangian theory with respect to the transformations

$$f(u) \to f(u) + a, \tag{8}$$

where a is arbitrary constant. This invariance means that all the space sections of manifold of system in question are equivalent.

Einstein himself put in correspondence to the gravity field symmetrical tensor field $\tilde{g}_{ij}$, which is characterized by the condition that adjoined quadratic differential form

$$d\tilde{s}^2 = \tilde{g}_{ij}du^i du^j, \tag{9}$$

has the signature of the interval in special relativity. In accordance with the principle of gravity physics discussed above, it is natural to assume that Einstein's interval (9) has a structure that is defined by the form-generating field g_{ij} (Riemann's metric (1)) and temporal field. If disclosed, this structure will give a simple method to introduce temporal field into the equations of gravitational physics. It is quite evident that the metric (1) has an Euclidean signature and hence it has no structure. To be transparent in our consideration, let us give a simple mathematical construction. One can consider tensor field S^i_j as linear transformation $\bar{V}^i = S^i_j V^j$ in the vector space in question. If the operator S is self-adjoint, that is $(V, SW) = (SV, W)$, then it is always possible to introduce the scalar product associated with operator S via the formula $<V,V>=(V,SV)$. It is clear that associated scalar product will be in general indefinite and with respect to the initial scalar product it has a structure. Thus, in general, the connection between the forms (1) and (9) is given by the relation $\tilde{g}_{ij} = g_{ik}S^k_j$. We shall give now simple expression for the operator S^i_j, which defines the Einstein interval. To this end we shall consider T-symmetry or time reversal invariance (7) as a fundamental physical principle. Further, we shall define T- symmetry in the space of the vector field in order to have transformation (7) as a particular case. We say that vector fields $\tilde{V}^i$ and V^i are T- symmetrical, if the sum of this fields is orthogonal to the gradient of temporal field and their difference is collinear to it, $(\tilde{V}^i + V^i)t_i = 0,\ \ \tilde{V}^i - V^i = \lambda t^i$. We have, $\tilde{V}^i = V^i - 2n^i(V,n) = (\delta^i_j - 2n^i n_j)V^j$, where

$$n^i = \frac{t^i}{\sqrt{(t,t)}}, \quad (t,t) = g_{ij}t^i t^j.$$

From this formula it follows that the fields t^i and $-t^i$ are T-symmetrical and hence the definition of T-symmetry given in the space of the vector field is correct. From the above consideration it follows that S is operator of T- symmetry that is $S^i_j = \delta^i_j - 2n^i n_j$ and hence for the Einstein's potential we obtain the following expression

$$\tilde{g}_{ij} = g_{ik}(\delta^k_j - 2n^k n_j) = g_{ij} - 2n_i n_j, \tag{10}$$

which is invariant under the transformations (7). The contravariant components of the tensor field $\tilde{g}_{ij}$ are $\tilde{g}^{ij} = g^{ij} - 2n^i n^j$, $\tilde{g}^{ik}\tilde{g}_{jk} = \delta^i_j$.

Let us give the physical meaning of the Einstein's scalar product associated with T- symmetry. Since

$$<V,V>=(V,V)-2(V,n)^2 = |V|^2(1-2\cos^2\phi) = -|V|^2\cos 2\phi,$$

where ϕ is the angle between the vectors V^i and n^i, the Einstein's scalar product is indefinite and can be positive, negative or equal to zero according to the value of the angle ϕ. In particular, $<V,V>=0$, if $\phi=\pi/4$. Thus, the Einstein's scalar product is time reversal invariant and permits to classify all the vectors depending on which angle they form with the gradient of the temporal field. As we see, the temporal field and T-symmetry define the Einstein's form (9) as the metric of the normal hyperbolic type. Hence, the gradient of the temporal field defines the causal structure on the physical manifold and can be identified with it. It is the physical meaning of the Einstein's interval and mechanism of formation of physical pseudo-Euclidian metric as well. It should be noted that in our reasoning we did not use the equation (5). This will become clear later under the consideration of the variational principle.

Now we shall show that the causal structure (the gradient of the temporal field) can be reduced to the canonical form $(0,0,0,1)$ by the suitable coordinate transformation at once in all points of some (may be small) patch of any point on the physical manifold. Local coordinates with respect to which gradient of the temporal field has the form $(0,0,0,1)$ will be called compatible with causal structure or intrinsic coordinates. It should be noted very important significance of the system of coordinates compatible with causal structure since it is similar to the Darboux system of coordinate in the theory of symplectic manifolds which is a geometrical basis for the Hamilton mechanics. What is more, there are transformations of coordinates which conserve the causal structure and these transformations are analogous to the canonical transformations in Hamiltonian mechanics.

Geometrically the stream of time is defined as a congruence of lines (lines of time) on the manifold. We recall that the congruence of lines is a set of lines characterized by the fact that only one element of the set passes through each point of manifold (or its open region). According to the definition, lines belonging to the congruence do not intersect and fill either the whole manifold or a part of it. The simple non-trivial example is the congruence of rays coming from one point of the Euclidean space.

Analytically the lines of time are defined as the solutions of the autonomous system of differential equations

$$\frac{du^i}{dt}=g^{ij}\frac{\partial f}{\partial u^j}=g^{ij}\partial_j f=(\nabla f)^i, \quad (i=1,2,3,4). \tag{11}$$

It can be shown that if the functions $\varphi^i(t)$ are solutions to Eqs. (11), the functions $\psi^i(t)=\varphi^i(t+a)$, where a is constant, will also be solutions to them. Since $ds^2=g_{ij}du^i du^j$, from Eqs. (5) and (11) it follows that $ds/dt=\pm 1$. Thus, the length of the time line is a linear function of the parameter t, $s=\pm t+a$. After these preliminary remarks, it is time to give solution of the above set problem.

Let

$$u^i(t)=\varphi^i(u_0^1,u_0^2,u_0^3,u_0^4,t)=\varphi^i(u_0,t) \tag{12}$$

be the solution to equations (11) with initial data $\varphi^i(u_0,t_0)=u_0^i$ so that

$$\partial\varphi^i(u_0,t_0)/\,\partial u_0^j=\delta^i_j.$$

Substituting $u^i(t)=\varphi^i(u_0,t)$ into the function $f(u^1,u^2,u^3,u^4)$ we obtain $p(t)=f(\varphi(u_0,t))$. Differentiating this function with respect to t, by virtue of (5) and (11), one finds

$$\frac{dp(t)}{dt}=\frac{\partial f}{\partial u^i}\frac{du^i}{dt}=g^{ij}\frac{\partial f}{\partial u^i}\frac{\partial f}{\partial u^j}=1.$$

It leads to

$$f(\varphi(u_0,t))=t-t_0+f(u_0).$$

Suppose that all initial data belong to the space section

$$f(u_0^1,u_0^2,u_0^3,u_0^4)=t_0.$$

Rewriting this relation in the parametric form

$$u_0^i=\psi^i(x^1,x^2,x^3),$$

Eqs. (12) can be written as the system of relations

$$u^i=\phi^i(x^1,x^2,x^3,t). \tag{13}$$

The functions (13) have continuous partial derivatives with respect to variable x^1,x^2,x^3,t and their functional determinant is not equal to zero. Hence the functions $\phi^i(x^1,x^2,x^3,t)$ design intrinsic system of coordinate of the dynamical system in question. Now one can show that in such a system of coordinates the covariant and contravariant components of the gradient of the temporal field and some components of the field g_{ij} take a simple numerical value

$$t^i=(0,0,0,1)=t_i,\quad g_{44}=g^{44}=1,\quad g_{\mu 4}=g^{\mu 4}=0,(\mu=1,2,3)$$

and hence this coordinate system is compatible with causal structure. Indeed, since in the system of coordinates x^1,x^2,x^3,t temporal field has a simple form

$$f(x^1,x^2,x^3,t)=f(\phi(x,t))=t,$$

then in this coordinate patch

$$\frac{\partial f}{\partial x^1} = \frac{\partial f}{\partial x^2} = \frac{\partial f}{\partial x^3} = 0, \quad \frac{\partial f}{\partial t} = 1$$

and hence $t_i = (0,0,0,1)$. Since coordinates x^1, x^2, x^3 do not vary along the lines of time, then

$$\frac{dx^\mu}{dt} = g^{\mu 4} = 0, \quad (\mu = 1,2,3) \quad \frac{dt}{dt} = g^{44} = 1.$$

Thus, in the intrinsic system of coordinates

$$g^{\mu 4} = 0, \quad g^{44} = 1, \quad t^i = g^{ij} t_j = g^{i4} = (0,0,0,1),$$

which is what we had to prove. Therefore, it follows that in the system of coordinates compatible with causal structure, metric (1) takes the form

$$ds^2 = g_{\mu\nu}(x^1, x^2, x^3, t) dx^\mu dx^\nu + (dt)^2, \quad \mu, \nu = 1,2,3, \tag{14}$$

since $t_i = g_{ij} t^j = g_{i4}$.

So, for any point of physical manifold we can indicate a local coordinate neighborhood with the coordinates compatible with causal structure. Covering manifold with such charts we get atlas on the manifold that is compatible with its causal structure. In this atlas the field equations have the most simple form in the sense that all components of the gradient of temporal field take numerical values. We see that causal structure is analogous to the symplectic structure of the Hamiltonian mechanics.

From the above consideration it also follows that variable t, parametrizing the line of time, can be considered as coordinate of time of the physical system in question. This name is justified particularly by the fact that the rate of change with time of any field is equal to the partial derivative with respect to t, i.e., $D_t = \partial/\partial t$ in the system of coordinate compatible with causal structure (see for example (6)). This is very important point since we see that the intrinsic coordinates give the possibility to write all equations in the form analogous to the canonical form of the Hamilton equations.

Furthermore, if we reverse time putting $\tilde{t}^i = -t^i$, then the lines of time will be parametrized by new variable $\tilde{t}$. From the equations (11) it is not difficult to derive that there is one-to-one and mutually continuous correspondence between the parameters t and $\tilde{t}$ given by the relation $\tilde{t} = -t$. From here it is clear that in the system of coordinate defined by the time reversal, the variable $-t$ will be the coordinate of time. Thus, the general covariant definition of time reversal given by (7) is adjusted with the familiar definition that is connected with the transformation of coordinates. Finally, we note one more important fact related to the intrinsic coordinates. Transformations (8) take the well-known form $t \to t + a$ in the system of coordinates compatible with the causal structure.

GEOMETRICAL LAWS OF THE GRAVITATIONAL FIELD

Consideration made above gives the evident geometrical method of constructing Lagrangians in gravity physics. It is clear that geometrical laws of the gravitational field are defined by the Lagrangian $L = \tilde{R}$, where $\tilde{R}$ is a scalar that is constructed from the Einstein's gravitational potential in the standard way.

Since the Einstein interval is defined by the two fields connected by the equation (5), it is necessary to pay special attention when deriving the equations of gravitation field. A standard method is to incorporate the constraint (5) via a Lagrange multiplier $\varepsilon = \varepsilon(u)$, rewrite the action density for gravity field in the form $L_g = \tilde{R} + \varepsilon(g^{ij}t_i t_j - 1)$ and treat the components of the fields g_{ij} and f as independent variables.

Let us consider the action

$$A = \frac{1}{2}\int \tilde{R}\sqrt{g}\, d^4u + \int L_m(\tilde{g}, F)\sqrt{g}\, d^4u + \frac{1}{2}\int \varepsilon(g^{ij}t_i t_j - 1)\sqrt{g}\, d^4u, \qquad (15)$$

where $g = Det(g_{ij}) > 0$ and $L_m(\tilde{g}, F)$ is the Lagrangian density of the system of other fields F which incorporates the Einstein's gravitational potential in the conventional form. Such geometrical method of introduction of causal structure into the geometrical equations of fields does not require special explanation but it is not evident how from geometrical laws to derive dynamical laws (the form of which was defined earlier). This problem will be solved here in the two important cases. It will be demonstrated how from the geometrical laws of the electromagnetic field to derive the Maxwell equations for the electric and magnetic fields in general covariant four-dimensional form (notice that this problem remains unresolved up-to-date). In other case we shall derive general covariant dynamical laws of the gravitational field from the geometrical laws of this field. In the framework of the questions discussed, now we would also like to mention that Riemann-Cartan geometry is defined by the connection

$$L^i_{jk} = \Gamma^i_{jk} + T^i_{jk} + g^{il}T^m_{lj}g_{mk} + g^{il}T^m_{lk}g_{mj},$$

where T^i_{jk} is the torsion tensor. Hence it is natural to incorporate the temporal field into this geometry by replacing g_{ij} on the Einstein gravitational potential $\tilde{g}_{ij}$.

Since Einstein's gravitational potential is the function of g_{ij} and f, it is necessary to use the chain rule. We have, $\delta\tilde{R} = \tilde{g}^{ij}\delta\tilde{R}_{ij} + \tilde{R}_{ij}\delta\tilde{g}^{ij}$. Further we denote the Christoffel symbols as Γ^i_{jk} and $\tilde{\Gamma}^i_{jk}$ belonging to the fields g_{ij} and $\tilde{g}_{ij}$, respectively. From (10), after some calculations, we obtain

$$\tilde{\Gamma}^i_{jk} = \Gamma^i_{jk} + H^i_{jk}, \qquad (16)$$

where

$$H^i_{jk} = n^i(\nabla_j n_k + \nabla_k n_j) + (2n^i n^l - g^{il})(n_j \delta^m_k + n_k \delta^m_j)(\nabla_m n_l - \nabla_l n_m).$$

In last formula ∇_j is covariant derivative with respect to the connection Γ^i_{jk}. It is easy to derive from (16) that $H^i_{ji} = 2n^i \nabla_j n_i = 0$, since $n^i n_i = 1$. Therefore, $\tilde{\Gamma}^j_{jk} = \Gamma^j_{jk}$, and hence $\tilde{\nabla}_l V^l = \nabla_l V^l$, where $\tilde{\nabla}_l$ is the covariant derivative with respect to the connection $\tilde{\Gamma}^i_{jk}$. From this it follows that $\tilde{g}^{ij}\delta\tilde{R}_{ij}$ can be omitted as a perfect differential. Varying now $\tilde{g}^{ij}$, we get

$$\delta\tilde{g}^{ij} = \delta g^{ij} + P^{ij}_{kl}\delta g^{kl} + Q^{ijk}\partial_k \delta f,$$

where

$$P^{ij}_{kl} = 2n^i n^j n_k n_l - n^i(n_k \delta^j_l + n_l \delta^j_k) - n^j(n_k \delta^i_l + n_l \delta^i_k),$$

$$Q^{ijk} = \frac{2}{\sqrt{(t,t)}}(2n^i n^j n^k - n^i g^{jk} - n^j g^{ik}).$$

A tensor field P^{ij}_{kl} is symmetrical in covariant and contravariant indices. Let $G_{ij} = \tilde{R}_{ij} - \frac{1}{2}\tilde{g}_{ij}\tilde{R}$ be the Einstein's tensor. Observing that

$$\tilde{g}_{ij} + \tilde{g}_{kl}P^{kl}_{ij} = g_{ij}, \quad g_{ij}Q^{ijk} = n_i n_j Q^{ijk} = 0,$$

it is easy to verify that variation $\delta(\tilde{R}\sqrt{g})$ can be presented in the form

$$\delta(\tilde{R}\sqrt{g}) = (G_{ij} + G_{kl}P^{kl}_{ij})\delta g^{ij}\sqrt{g} - \nabla_k(G_{ij}Q^{ijk})\delta f\sqrt{g}$$

with neglect of a perfect differential. Further we put $\delta L_F = \frac{1}{2}M_{ij}\delta\tilde{g}^{ij}$, and introduce by the standard way the energy-momentum tensor $T_{ij} = M_{ij} - \tilde{g}_{ij}L_F$. Then

$$\delta(L_F\sqrt{g}) = \frac{1}{2}(T_{ij} + T_{kl}P^{kl}_{ij})\delta g^{ij}\sqrt{g} - \nabla_k(T_{ij}Q^{ijk})\delta f\sqrt{g}$$

with neglect of a perfect differential. Thus a total variation of the action can be presented in the following form

$$\delta A = \frac{1}{2}\int (A_{ij})\delta g^{ij} + B\delta f + \delta\varepsilon(g^{ij}t_i t_j - 1))\sqrt{g}d^4u, \tag{17}$$

where

$$A_{ij} = G_{ij} + G_{kl}P^{kl}_{ij} + T_{ij} + T_{kl}P^{kl}_{ij} + \varepsilon t_i t_j - \varepsilon(g^{kl}t_k t_l - 1)g_{ij}),$$

$$B = -\nabla_k((G_{ij} + T_{ij} - \varepsilon t_i t_j)Q^{ijk}) - 2\nabla_k(\varepsilon t^k).$$

One can consider tensor P^{kl}_{ij} as operator P acting in the space of symmetrical tensor fields. The characteristic equation of this operator has the form $P^2 + 2P = 0,$ and hence $(P+1)^2 = 1$. Thus, operator $P+1$ is inverse to itself. Since $t_i t_j + t_k t_l P^{kl}_{ij} = -t_i t_j$, from (17) it follows that in framework of considered here concept of time the Einstein equations have the form

$$G_{ij} + T_{ij} = \varepsilon\partial_i f \partial_j f, \quad g^{ij}\partial_i f \partial_j f = 1, \tag{18}$$

$$\nabla_k(\varepsilon t^k) = 0. \tag{19}$$

The equations (18) constitute the full system of geometrical equations of the gravitational field. As it is shown above, these equations emerge from the first principles of gravity physics formulated by Einstein.

Eq. (19) expresses the law of energy conservation in gravitational physics which, evidently, is general covariant. To make sure that we indeed deal with conservation of energy, it is sufficient to figure out that action (15) is invariant with respect to transformation (8). Thus, the equation (19) results also from the Noether's theorem and the invariance of the action with respect to the one parametric group of transformations (8) and it should be derived from Eqs. (18). It is also clear that the Lagrange multiplier ε has a physical meaning of energy density of the system in question. From the Eqs. (18) it follows that

$$\varepsilon = G_{ij}t^i t^j + T_{ij}t^i t^j = \varepsilon_g + \varepsilon_m, \tag{20}$$

where $\varepsilon_g = G_{ij}t^i t^j = \frac{1}{2}\tilde{R}_{ij}g^{ij}$ is energy density of the gravitational field and $\varepsilon_m = T_{ij}t^i t^j$ is energy density of other fields. Consider the law of energy conservation from the various points of view. First of all we consider link between (18) and (19). The so-called local energy conservation is written as follows $\tilde{\nabla}_i T^{ij} = 0$, where $T^{ij} = T_{kl}\tilde{g}^{ik}\tilde{g}^{jl}$. These equations are fulfilled on the equations of the fields F that contribute to the energy-momentum tensor. Since $\tilde{\nabla}_i G^{ij} = 0$ identically, from the Einstein's field equations (18) it follows that

$$\tilde{\nabla}_i T^{ij} = \varepsilon t^i(\tilde{\nabla}_i t^j) + t^j\tilde{\nabla}_i(\varepsilon t^i).$$

We now show that $t^i\tilde{\nabla}_i t^j = 0$. From (5) we have $n_i = t_i$, $\nabla_i n_j - \nabla_j n_i = \nabla_i t_j - \nabla_j t_i = 0$. With this and from (16), we get $\tilde{\Gamma}^i_{jk} = \Gamma^i_{jk} + 2t^i\nabla_j t_k$. Thus, $t^i\tilde{\nabla}_i t^j = t^i\nabla_i t^j + 2t^i t^j t^l \nabla_i t_l = 0$. Since $\tilde{\nabla}_i(\varepsilon t^i) = \nabla_i(\varepsilon t^i)$, finally we have

$$\tilde{\nabla}_i T^{ij} = t^j \nabla_i (\varepsilon t^i).$$

In view of this the energy conservation law can be treated as the condition of compatibility of the field equations. In this sense, the law of energy conservation is analogous to the law of charge conservation.

Show that rate of change with time of the energy density $D_t(\sqrt{g}\varepsilon)$ equal to zero $D_t(\sqrt{g}\varepsilon) = 0$ and hence this quantity is a first integral of the system in question. We have $D_t(\sqrt{g}\varepsilon) = t^i\partial_i(\sqrt{g}\varepsilon) + \sqrt{g}\varepsilon\partial_i t^i = \partial_i(\sqrt{g}\varepsilon t^i)$. Since $\sqrt{g}\nabla_k(\varepsilon t^k) = \partial_i(\sqrt{g}\varepsilon t^i)$, from the law of energy conservation (19) it follows that energy density is the first integral of the system

$$D_t(\sqrt{g}\varepsilon) = 0.$$

In the system of coordinates, compatible with causal structure, this equation has a more customary form

$$\frac{\partial}{\partial t}(\sqrt{g}\varepsilon) = 0.$$

Compare the law of energy conservation with the law of charge conservation which can be written in general covariant form as follows $\nabla_i J^i = 0$. Putting $J^i = t^i(t,J) + J^i - t^i(t,J) = \rho t^i + P^i$, we find that $D_t(\sqrt{g}\rho) + \partial_i(\sqrt{g}P^i) = 0$. Thus, we see that the charge density is not in general the first integral of the system while energy density always is.

The Eqs. (18) are geometrical and from the physical point of view it is not evident that they express the fundamental dynamical laws of the nature. This is not a simple exercise to derive from (18) a physical laws of the gravitational field and we first consider more easy (but very important as well) problem. We derive the general covariant Maxwell equations from some natural geometrical equations.

TIME IN THE THEORY OF ELECTRIC AND MAGNETIC FIELDS

Let A_i be the vector potential of the electromagnetic field. Let us define the gauge invariant tensor of electromagnetic field as usual $F_{ij} = \partial_i A_j - \partial_j A_i$. According to the principle of gravity physics we introduce temporal field into the theory of electromagnetic field through the gauge invariant and general covariant Lagrangian

$$L_{em} = \frac{1}{4} F_{ij} F_{kl} \tilde{g}^{ik} \tilde{g}^{jl}.$$

with the energy-momentum tensor

$$T_{ij} = F_{ik} F_{jl} \tilde{g}^{kl} - \tilde{g}_{ij} L_{em}. \tag{21}$$

The equations for the tensor F_{ij} can be written in the form

$$\nabla_i \overset{*}{F}{}^{ij} = 0, \quad \nabla_i \tilde{F}^{ij} = 0, \tag{22}$$

where

$$\overset{*}{F}{}^{ij} = \frac{1}{2} e^{ijkl} F_{kl}, \quad \tilde{F}^{ij} = F_{kl} \tilde{g}^{ik} \tilde{g}^{jl},$$

and e^{ijkl} are contravariant components of the Levi-Civita tensor normalized as $e_{1234} = \sqrt{g}$. Now we shall derive from geometrical laws of the electromagnetic field the general covariant four dimensional Maxwell equations for the electric and magnetic fields which in fact express the fundamental physical laws. First of all we formulate the main relations of the vector algebra and vector analysis on the four dimensional physical manifold which is interesting by itself.

A scalar product of two vector fields A^i and B^i is defined by the Riemann metric (1) as earlier. A vector product of two vector fields A^i and B^i we shall construct as follows

$$C^i = [AB]^i = e^{ijkl} t_j A_k B_l.$$

Here the crucial significance of the temporal field should be emphasized. It is evident that $[AB] + [BA] = 0$. From geometrical point of view a vector product is a vector field that is tangent to any space section of the manifold in any point. By the direct calculation it can be shown that $|[AB]| = |A||B| \sin\varphi$, $[A[BC]] = B(A,C) - C(A,B)$. It should be noted that vector product is invariant with respect to transformations

$$A_i \to A_i + \lambda t_i, \quad B_i \to B_i + \mu t_i,$$

where λ and μ are scalar fields.

Differential operators of the vector analysis on the physical manifold are defined as natural as algebraic ones. For the divergence and gradient we have respectively

$$div\, A = \nabla_i A^i = \frac{1}{\sqrt{g}} \partial_i (\sqrt{g} A^i),$$

$$(grad\, \phi)^i = (g^{ij} - t^i t^j) \partial_j \phi = g^{ij} \partial_j \phi - t^i D_t \phi$$

and as a consequence

$$div\, grad\phi = \frac{1}{\sqrt{g}}\partial_i(\sqrt{g}g^{ij}\partial_j\phi) - \frac{1}{\sqrt{g}}\partial_i(\sqrt{g}t^i D_t\phi) = \nabla_i\nabla^i\phi - \nabla_i(t^i D_t\phi).$$

A rotor of the vector field A is defined as a vector product of ∂ and A

$$(rot\ A)^i = e^{ijkl}t_j\partial_k A_l = \frac{1}{2}e^{ijkl}t_j(\partial_k A_l - \partial_l A_k).$$

It is easy to verify that

$$rot\ grad\,\phi = 0.$$

Since the temporal field satisfies the equations $\partial_i t_j - \partial_j t_i = 0$, then rotor is invariant with respect to the transformations $A_i \to A_i + \lambda t_i$, where λ is a scalar field. It is evident that all operations so defined are general covariant.

There is only one direct way to derive from the geometrical Eqs. (22) the fundamental physical laws first formulated by Maxwell. Consider the rate of change with time of the potential of electromagnetic field. By definition, we have

$$D_t A_i = t^k\partial_k A_i + A_k\partial_i t^k = t^k(\partial_k A_i - \partial_i A_k) + \partial_i(t^k A_k) = t^k F_{ki} - \partial_i\phi,$$

where $\phi = -t^k A_k$. Thus, the rate of change with time of the electromagnetic potential can be presented as the difference of two covector fields with one of them having the form

$$E_i = t^k F_{ik}. \tag{23}$$

From this it follows that strength of the electric field is general covariant and gauge invariant quantity that is defined by the equation

$$E_i = t^k F_{ik} = -D_t A_i - \partial_i\phi.$$

Now it is quite clear that general covariant and gauge invariant definition of the magnetic field strength is given by the formula

$$H^i = (rotA)^i = e^{ijkl}t_j\partial_k A_l = \frac{1}{2}e^{ijkl}t_j(\partial_k A_l - \partial_l A_k).$$

Thus,

$$H_i = t^k \overset{*}{F}_{ik}, \tag{24}$$

where $\overset{*}{F}_{ij} = g_{ik}g_{jl}\overset{*}{F}{}^{kl}$. It is evident that vectors E and H are orthogonal to gradient of temporal field

$$t^i H_i = t^i E_i = 0.$$

Thus, vectors E^i and H^i belong to the three dimensional linear space of the vector fields orthogonal to the gradient of the temporal field.

Now it is not difficult to derive the Maxwell equations for the electric and magnetic fields staying in the framework of the vector analysis. However, our goal is to test the new concept of time and hence to give derivation of the fundamental equations of physics precisely from the geometrical Eqs. (22).

Resolving equations (23), (24) over F_{ik}, we obtain

$$F_{ij} = -t_i E_j + t_j E_i - \varepsilon_{ijkl} t^k H^l. \tag{25}$$

Thus, on the physical manifold there is general covariant one-to-one algebraic relation between the electric and magnetic fields and tensor of the electromagnetic field that is defined by the temporal field. Out of (25) we find

$$\overset{*}{F}{}^{ij} = -t^i H^j + t^j H^i - e^{ijkl} t_k E_l, \quad \tilde{F}^{ij} = t^i E^j - t^j E^i - e^{ijkl} t_k H_l. \tag{26}$$

Substituting (26) into (22), we shall obtain the Maxwell equations for the strengths of the electric and magnetic fields in the following general covariant and gauge invariant form

$$D_t H^i + H^i \nabla_k t^k = -e^{ijkl} t_j \nabla_k E_l, \quad div\, H = \nabla_i H^i = \frac{1}{\sqrt{g}} \partial_i (\sqrt{g} H^i) = 0$$

$$D_t E^i + E^i \nabla_k t^k = e^{ijkl} t_j \nabla_k H_l, \quad div\, E = \nabla_i E^i = \frac{1}{\sqrt{g}} \partial_i (\sqrt{g} E^i) = 0.$$

Derive, for example, first set of the Maxwell equations. With (26) we get

$$-\nabla_i \overset{*}{F}{}^{ij} = \nabla_i (t^i H^j - t^j H^i + e^{ijkl} t_k E_l) =$$
$$= t^i \nabla_i H^j - H^i \nabla_i t^j + H^j \nabla_i t^i - t^j \nabla_i H^i + e^{jkil} t_k \nabla_i E_l.$$

Since,

$$D_t H^j = t^i \partial_i H^j - H^i \partial_i t^j = t^i \nabla_i H^j - H^i \nabla_i t^j,$$

we have

$$-\nabla_i \overset{*}{F}{}^{ij} = D_t H^j + e^{jkil} t_k \nabla_i E_l + H^j \nabla_i t^i - t^j \nabla_i H^i.$$

Taking into account that $D_t t_i = D_t t^i = 0, \quad t_i H^i = 0,$ we derive from the last formula what we need.

Since $\sqrt{g}\nabla_i t^i = \partial_i(\sqrt{g}t^i)$, $D_t\sqrt{g} = t^i\partial_i\sqrt{g} + \sqrt{g}\partial_i t^i = \partial_i(\sqrt{g}t^i)$, the dynamical equations of the electromagnetic field may be written in more compact form

$$\frac{1}{\sqrt{g}}D_t(\sqrt{g}H) = -rotE, \quad \frac{1}{\sqrt{g}}D_t(\sqrt{g}E) = rotH. \tag{27}$$

From Eqs. (27) it follows that

$$D_t(\partial_i(\sqrt{g}H^i)) = 0, \quad D_t(\partial_i(\sqrt{g}E^i)) = 0,$$

and therefore the dynamical equations are compatible with constraints. From Eqs. (27) one can also derive that

$$D_t(\sqrt{g}t_iE^i) = 0, \quad D_t(\sqrt{g}t_iH^i) = 0,$$

which shows that the orthogonality of electric and magnetic fields to the gradient of the temporal field may be considered as the initial conditions.

Thus, it is shown that the principles of the theory of time and gravity physics are in full correspondence with the fundamental physical laws and hence they can be considered as a method to derive new fundamental equations. To complete this discussion with the Maxwell equations as the main topic, we should only note that the second set of the Maxwell equations can be derived from the principle of least action with the Lagrangian

$$L_{em} = \frac{1}{2}(|E|^2 - |H|^2).$$

We also write the expression for the components of the energy-momentum tensor in terms of electric and magnetic field strength

$$T_{ij} = \frac{1}{2}g_{ij}(|E|^2 + |H|^2) - E_iE_j - H_iH_j - t_i\Pi_j - t_j\Pi_i, \tag{28}$$

where Π_i are covariant components of the Pointing vector

$$\Pi^i = e^{ijkl}t_jE_kH_l, \quad \Pi = [EH].$$

From (28) and definition given above we find the energy density of the electromagnetic field

$$\varepsilon_{em} = \frac{1}{2}(E^2 + H^2)$$

which is a reasonable result.

For the completeness of the picture we shall formulate the energy conservation law starting from the general covariant Maxwell equations. Out of (27) we immediately get

$$E_l D_t E^l + H_l D_t H^l + (E^2 + H^2)\nabla_i t^i + \nabla_i \Pi^i = 0.$$

Taking into account the relations

$$\frac{1}{2} D_t (E_l E^l) = \frac{1}{2} D_t \mid E \mid^2 = E_l D_t E^l + \frac{1}{2} E^l E^k D_t g_{lk},$$

$$\nabla_i t^i = \frac{1}{2} g^{ij} (\nabla_i t_j + \nabla_j t_i) = \frac{1}{2} g^{ij} D_t g_{ij} = \frac{1}{\sqrt{g}} D_t \sqrt{g},$$

this equation may be written in the form

$$\frac{1}{\sqrt{g}} D_t (\sqrt{g} \varepsilon_{em}) + \nabla_i \Pi^i = -\frac{1}{2} T^{ij} D_t g_{ij}, \tag{29}$$

where $T^{ij} = \tilde{g}^{ik} \tilde{g}^{jl} T_{kl}$. Let us now assume that the electromagnetic field is considered on the background of the physical manifold of some full system of fields. Let it be known also that for gravity field of this system the equation $D_t g_{ij} = 0$ holds valid. In such approximation, when physical manifold is external with respect to the electromagnetic field, from (29) we obtain that the energy density of the electromagnetic field satisfies the equation

$$\frac{1}{\sqrt{g}} D_t (\sqrt{g} \varepsilon_m) + \nabla_i \Pi^i = 0.$$

It is exactly the energy conservation law of the electromagnetic field in the above mentioned approximation.

To conclude this section we mention the existence of the Lagrangian

$$L_{CS} = \frac{q}{2} t_i A_j \overset{*}{F}{}^{ij}$$

that is not invariant with respect to time reversal. With this, the equations (27) take the following form

$$\frac{1}{\sqrt{g}} D_t (\sqrt{g} H) = -rotE, \quad \frac{1}{\sqrt{g}} D_t (\sqrt{g} E) = rotH + qH.$$

The modification of electrodynamics due to the Lagrangian of this form was considered in [11] where an analogue of the gradient of the temporal field was introduced and interpreted as a mass of the photon. It would be very interesting to reconsider the results of the paper [11] from the new point of view presented here.

PHYSICAL LAWS OF THE GRAVITATIONAL FIELD

Now we have an important experience that allows us to write geometrical laws (18) in the form of physical equations of the gravitational field analogous to the Maxwell equations. The guiding idea is very simple. The equation $div\, E = 4\pi\rho$ is tightly connected with the charge conservation. Thus, our goal is to find gravitational analog of this equation which should be connected with the energy conservation in a manner similar to the conservation of the electric charge.

To this end we first of all introduce the important notion of the momentum of gravitational field, i.e. quantity which is analogous to $E_i = D_t A_i + \partial_i (t^i A_i)$. Consider a tensor field of the type (1,1)

$$P^i_j = \frac{1}{2} g^{ik} D_t g_{jk} = \frac{1}{2} g^{ik} (\nabla_j t_k + \nabla_k t_j) = g^{ik} \nabla_j t_k = \nabla_j t^i .$$

A tensor field so defined will be called the momentum of the gravitational field or simply the momentum of the field. Give some important relations for the momentum

$$t_i P^i_j = 0, \quad t^j P^i_j = 0, \quad g^{ik} P^l_k g_{lj} = P^i_j .$$

$$D_t P^i_j = t^l \nabla_l P^i_j, \quad \nabla_i P^k_j - \nabla_j P^k_i = R_{ijl}{}^k t^l .$$

$$D_t \Gamma^i_{jk} = \nabla_j P^i_k + \nabla_k P^i_j - g^{il} g_{km} \nabla_l P^m_j, \quad D_t g = g g^{ik} D_t g_{ik} = 2g P^i_i .$$

As the following step in the required direction we shall write equations (18) in other form.

By contraction with $\tilde{g}^{ij}$ we get from (18) that $\tilde{R} = \varepsilon + T$, where $T = T_{ij} \tilde{g}^{ij}$. Using this relation we transform (18) to the following form

$$\tilde{R}_{ij} + T_{ij} - \frac{1}{2} \tilde{g}_{ij} T = \frac{1}{2} g_{ij}\, \varepsilon. \tag{30}$$

If we start with equations (30), we get by contraction (20) and $\tilde{R} = \varepsilon + T$, and so we can get back to (18). We may use either (18) or (30) as the basic equations.

With (5) and (16) we get $\tilde{R}_{ij} = R_{ij} + \nabla_l (t^l D_t g_{ij})$ and hence $g^{ik} \tilde{R}_{jk} = R^i_j + 2 D_t P^i_j + 2 P^k_k P^i_j$. Introduce the tensor fields

$$B^i_j = h^i_k R^k_l h^l_j + D_t P^i_j + P^k_k P^i_j ,$$

$$N^i_j = h^i_k T^k_l h^l_j - T h^i_j ,$$

where $h^i_j = \delta^i_j - t^i t_j$ is projection operator, $h^i_k h^k_j = h^i_j$. We next define the flux vector of the energy of the gravitational field

$$G_i = \nabla_k P^k_i - \partial_i P^k_k + t_i (P^k_l P^l_k + D_t P^k_k)$$

and the flux vector of the energy of other fields

$$\Pi_i = \varepsilon_m t_i - T_{ik} t^k .$$

It is evident that $(t, G) = (t, \Pi) = 0$. With this we can derive from the equations (30) the following system of equations

$$D_t P^i_j + P^k_k P^i_j + B^i_j + N^i_j = \frac{1}{2} \varepsilon h^i_j \tag{31}$$

$$G_i = \Pi_i \tag{32}$$

and vice versa. Our statement is that equations (31) represent in general covariant form the physical laws of the gravitational field similar to the Maxwell equations (27). To emphasize this analogy we write equations (31) in the form

$$\frac{1}{\sqrt{g}} D_t (\sqrt{g} P^i_j) + B^i_j + N^i_j = \frac{1}{2} \varepsilon h^i_j$$

and add that like equation $divE = 0$, the equation (32) may be considered as constraint since from (31) it can be derived that

$$D_t (G_i - \Pi_i) + P^k_k (G_i - \Pi_i) = \frac{1}{\sqrt{g}} D_t (\sqrt{g} (G_i - \Pi_i)) = 0.$$

The last equation means also that equations (31) and (32) are compatible.

Equation (32) has very simple physical sense that the flux vector of the energy of the gravitational field is exactly equal to the flux vector of energy of other fields . For the energy density of the gravitational field we have $\varepsilon_g = \frac{1}{2} \tilde{R}_{ij} g^{ij} = \frac{1}{2} R + (P^i_i)^2 + D_t P^i_i = T + U$, where

$$T = \frac{1}{2} (P^i_i)^2 - \frac{1}{2} P^i_j P^j_i \tag{33}$$

is a density of kinetic energy and

$$U = \frac{1}{2} R + \frac{1}{2} (P^i_i)^2 + \frac{1}{2} P^i_j P^j_i + D_t P^i_i \tag{34}$$

is a density of potential energy of the gravitational field.

For the better understanding of the last constructions we shall consider the general covariant dynamical laws of the gravitational field in the atlas compatible with causal structure. With respect to the coordinate transformations that conserve causal structure the quantity $g_{\mu\nu}(t,x^1,x^2,x^3)$, $(\mu,\nu=1,2,3)$ from the metric (14) transforms as the symmetrical tensor and $dl^2 = g_{\mu\nu}(t,x^1,x^2,x^3)dx^\mu dx^\nu$ may be considered as the metrics of the space sections of the physical manifold. For this one parametric set of three dimensional metrics we can consider Christoffel symbols $L^\mu_{\nu\sigma}$ and covariant derivative ∇_μ, Riemannian tensor curvature $S_{\mu\nu\sigma}{}^\tau$, tensor Ricci $S_{\nu\sigma} = S_{\mu\nu\sigma}{}^\mu$ and scalar curvature $S = g^{\mu\nu}S_{\mu\nu} = S^\mu_\mu$ as usual. In the atlas compatible with causal structure we have $L^\mu_{\nu\sigma} = \Gamma^\mu_{\nu\sigma}$, $P^4_i = P^i_4 = 0$, $B^4_i = B^i_4 = 0, (i = 1,2,3,4)$ and what is more important that

$$B^\mu_\nu = S^\mu_\nu, \quad R + (P^i_i)^2 + P^i_j P^j_i + 2D_t P^i_i = S$$

Thus, out of (33), (34) we get for the densities of the kinetic and potential energy of the gravitational field following representation

$$T = \frac{1}{2}(P^\sigma_\sigma)^2 - \frac{1}{2}P^\mu_\nu P^\nu_\mu, \quad U = \frac{1}{2}S.$$

Now we write equations (31) and (32) in the atlas compatible with causal structure under assumption that T_{ij} is the tensor of the energy–momentum of the electromagnetic field. In this case

$$N^\mu_\nu = T^\mu_\nu = \frac{1}{2}(E^2 + H^2)\delta^\mu_\nu - E^\mu E_\nu - H^\mu H_\nu$$

is the Maxwell stress energy tensor. Thus, we can write the basic equations (31) and (32) in the following form

$$\dot{P}^\mu_\nu + P^\sigma_\sigma P^\mu_\nu + S^\mu_\nu + T^\mu_\nu = \frac{1}{2}\varepsilon\delta^\mu_\nu, \tag{35}$$

$$\nabla_\nu P^\nu_\mu - \partial_\mu P^\sigma_\sigma = \Pi_\mu = [\mathbf{EH}]_\mu, \tag{36}$$

where dot stands for partial derivative with respect the variable t. From equations (35) and (36) it follows that the Cauchy problem for the gravitational field is not more difficult than for the electromagnetic field.

In the above consideration we derived the law of energy conservation from very general geometrical laws. It is very important from the physical point of view to consider energy conservation on the basis of Maxwell equations and physical laws of gravitational field (35) and (36). In the atlas compatible with causal structure the Maxwell equations (27) read

$$\dot{E}^{\mu} + P^{\sigma}_{\sigma} E^{\mu} = e^{\mu\nu\sigma} \partial_{\nu} H_{\sigma}, \quad \dot{H}^{\mu} + P^{\sigma}_{\sigma} H^{\mu} = -- e^{\mu\nu\sigma} \partial_{\nu} E_{\sigma}.$$

Since $\dot{g} = g g^{\mu\nu} g_{\mu\nu} = 2gP^{\sigma}_{\sigma}$,

$$\frac{\partial}{\partial t}(\sqrt{g}\,\varepsilon) = \sqrt{g}(\dot{\varepsilon} + P^{\sigma}_{\sigma}\,\varepsilon).$$

Taking into account that $\dot{g}_{\mu\nu} = 2g_{\nu\sigma} P^{\sigma}_{\mu}$ one can derive from the Maxwell equations the formula

$$\dot{\varepsilon}_{em} + P^{\sigma}_{\sigma}\,\varepsilon_{em} = -\nabla_{\nu}\Pi^{\nu} - T^{\mu}_{\nu} P^{\nu}_{\mu}.$$

In the atlas compatible with causal structure we have for the energy density of the gravitational field

$$\dot{\varepsilon}_g = P^{\sigma}_{\sigma}\,\dot{P}^{\sigma}_{\sigma} - P^{\nu}_{\mu}\,\dot{P}^{\mu}_{\nu} + \frac{1}{2}\dot{S}.$$

From the equations (35) we have the relations

$$\dot{P}^{\sigma}_{\sigma} = \frac{1}{2}\varepsilon_{em} - \frac{1}{2}\varepsilon_{g} - P^{\nu}_{\mu} P^{\mu}_{\nu}, \quad P^{\nu}_{\mu}\,\dot{P}^{\mu}_{\nu} = \frac{1}{2}\varepsilon P^{\sigma}_{\sigma} - P^{\sigma}_{\sigma} P^{\nu}_{\mu} P^{\mu}_{\nu} - P^{\nu}_{\mu} S^{\mu}_{\nu} - P^{\nu}_{\mu} T^{\mu}_{\nu}.$$

With this one can show that

$$\dot{\varepsilon}_g + P^{\sigma}_{\sigma}\,\varepsilon_g = \frac{1}{2} g^{\mu\nu}\,\dot{S}_{\mu\nu} + T^{\mu}_{\nu} P^{\nu}_{\mu}.$$

Since $\frac{1}{2} g^{\mu\nu}\,\dot{S}_{\mu\nu} = \nabla_{\nu} G^{\nu}$, for $\varepsilon = \varepsilon_g + \varepsilon_{em}$ we have

$$\dot{\varepsilon} + P^{\sigma}_{\sigma}\,\varepsilon = \nabla_{\nu} G^{\nu} - \nabla_{\nu}\Pi^{\nu}.$$

Thus,

$$\frac{\partial}{\partial t}(\sqrt{g}\varepsilon) = \partial_{\nu}(\sqrt{g} G^{\nu}) - \partial_{\nu}(\sqrt{g}\Pi^{\nu}). \tag{37}$$

We may integrate this relation over a three-dimensional volume V lying in the space section of the physical manifold. For the energy in the volume V we have $W = \int \varepsilon \sqrt{g} dV$. Consider the fluxes of the gravitational $\Phi_g = \int \Omega_{\nu} d\sigma^{\nu}$ and the electromagnetic $\Phi_{em} = \int \Pi_{\nu} d\sigma^{\nu}$ energy, where integral is taken over the boundary surface of the volume

V. Since the right hand side of the equation (37) can be converted by Gauss's theorem, from (37) we have the relation

$$\frac{\partial}{\partial t}W = \Phi_g - \Phi_{em}.$$

Thus, the law of energy conservation means that the fluxes of the gravitational and electromagnetic energy flow in opposite directions and exactly equal to each other. (We know that our sun is a surprisingly powerful source of the radiant energy. Now it is evident that if there is the flow of gravitational energy in the direction to the sun then we can say that our sun is the transducer of the gravitational energy into the electromagnetic one.)

It is useful to show that equations (35) and (36) are compatible. Indeed, from equations (35) it follows that

$$\dot{U}_\mu + P^\sigma_\sigma U_\mu = \frac{1}{\sqrt{g}}\frac{\partial}{\partial t}(\sqrt{g}U_\mu) = 0,$$

where $U_\mu = G_\mu - \Pi_\mu$. To derive this equation the following formula should be used

$$\dot{\Gamma}^\sigma_{\mu\nu} = \nabla_\mu P^\sigma_\nu + \nabla_\nu P^\sigma_\mu - g^{\sigma\lambda} g_{\nu\tau} \nabla_\lambda P^\tau_\mu .$$

Now we consider question about exact solutions of the equations (35) and (36).

EXACT SOLUTIONS

We have seen above that there is deep analogy between the energy and charge and the gravitational and electromagnetic fields. Here we consider the existence of solutions to the equations (35) and (36) which reproduce the physical situation with pointlike distribution of the gravitational energy, i.e. $\varepsilon_g = m\delta(r)$, where m is a gravitational charge of point source. This physical situation is analogous to the pointlike distribution of density of charge, when $\rho = q\delta(r)$, where q is a electric charge of point source. Thus, we search the solution of the gravitational equations analogous to the Coulomb potential. Starting point is the question about static gravitational field.

Gravitational fields will be called **static** if the rate of change with time of the gravitational potential is equal to zero, $D_t g_{ij} = 0$. In accordance with (6), this condition can be written as

$$D_t g_{ij} = \nabla_i t_j + \nabla_j t_i = 0.$$

It is evident that definition of the static gravitational fields is general covariant. Since $\nabla_i t_j - \nabla_j t_i = 0$, then for a static gravitational field we have $\nabla_i t_j = 0$ and hence $R_{ijk}{}^l t_l = 0$, $\tilde{R}_{ijk}{}^l = R_{ijk}{}^l$. With this it is not difficult to show that a static gravitational field is absent ($R_{ijk}^{\ \ \ l} = 0.$) Thus, the gravitational field generated by the gravitational charge can not be static.

The following point is the definition of the spherical symmetrical gravitational potential. We use the empirical (non internal) representations about the spherical symmetry. In accordance with this in the atlas compatible with causal structure a potential with spherical symmetry has the form

$$dl^2 = A^2 dr^2 + B^2(d\theta^2 + \sin^2\theta\, d\phi^2),$$

where $A = A(r,t)$, $B = B(r,t)$. For the nonzero components of the Ricci tensor of this one parametric set of metric we find the following expressions

$$S_{11} = \frac{2}{AB}(A'B' - AB''), \quad S_{22} = \frac{B}{A^3}(A'B' - AB'') + 1 - \frac{B'^2}{A^2}$$

and $S_{33} = S_{22}\sin^2\theta$, where prime stands for derivative with respect to the variable r. Hence, for the potential energy of this configuration we have

$$U = \frac{1}{2}S = \frac{2}{A^3 B}(A'B' - AB'') + \frac{1}{B^2}\left(1 - \frac{B'^2}{A^2}\right).$$

To get $U = 0$ we immediately put $B' = A$. Remind, that we aim to realize the situation with a pointlike distribution of the gravitational energy.

For the nontrivial components of the momentum we have

$$P_1^1 = \frac{\dot{A}}{A}, \quad P_2^2 = P_3^3 = \frac{\dot{B}}{B},$$

where dot denotes partial derivative with respect to the variable t. The flux vector of the energy has only one nonzero component

$$G_1 = \frac{2}{AB}(\dot{A}B' - A\dot{B}')$$

and in accordance with relation $A = B'$ the equation $G_i = 0$ is fulfilled. Thus, we have two equations

$$2B\ddot{B} + \dot{B}^2 = 0,$$

$$\ddot{A}B^2 + \dot{A}\dot{B}B = \frac{1}{2}A\dot{B}^2.$$

Now we use the other empirical idea about a spherical wave and put

$$A = A(r - t), \quad B = B(r - t).$$

From the equation for B only we find that $B\dot{B}^2 = m,$ where m is a constant of integration (gravitational charge) and hence

$$B = m\left[\frac{3}{2}\frac{(r-t)}{m}\right]^{\frac{2}{3}}.$$

Since $A = B^{'}$, then

$$A^{-1} = \left[\frac{3}{2}\frac{(r-t)}{m}\right]^{\frac{1}{3}}$$

and it is not difficult to see that second equation for A and B is fulfilled automatically. Thus, the Einstein gravitational potential can be presented in the following form

$$d\tilde{s}^2 = -dt^2 + \frac{dr^2}{\left[\frac{3}{2}\frac{(r-t)}{m}\right]^{\frac{2}{3}}} + \left[\frac{3}{2}\frac{(r-t)}{m}\right]^{\frac{4}{3}} m^2(d\theta^2 + \sin^2\theta d\phi^2).$$

The solution obtained is known as the Schwarzschild solution [12]. Since for the kinetic energy we have $T = 0,$ then we see that this solution really correspond to the situation described above and hence in definite sense it is an analogous of the Coulomb potential.

Now we consider another very simple but important physical situation. The idea about homogeneous and isotropic distribution of the gravitational energy may be realized in the atlas compatible with causal structure by the one parametric set of metric

$$dl^2 = a(t)d\sigma^2,$$

where $d\sigma^2$ is the metric of the unit 3d sphere, which in the four dimensional spherical coordinates has the form

$$d\sigma^2 = d\psi^2 + \sin^2\psi(d\theta^2 + \sin^2\theta d\phi^2).$$

Now problem is to find a function $a(t).$ We have

$$S^\mu_\nu = \frac{2}{a^2}\delta^\mu_\nu, \quad P^\mu_\nu = \frac{\dot{a}}{a}\delta^\mu_\nu$$

and hence for the density of the gravitational energy we get

$$\varepsilon_g = \frac{3}{a}(\dot{a}^2 + 1).$$

We know that $\partial(\sqrt{g}\,\varepsilon_g)/\partial t = 0$. Since $\sqrt{g} = a^3 \sin^2\psi \sin\theta$ then by integrating over the variables ψ, θ, ϕ we get the following equation for $a(t)$ from the law of energy conservation

$$a(\dot{a}^2 + 1) = 2a_0 = const.$$

With respect to the new variable η, such that $dt = ad\eta$ the solution of the equation in question can be presented in the following form $a = a_0(1 - \cos\eta)$. Thus, we get that emergence and evolution of the clot of gravitational energy is described in parametric form as follows

$$a = a_0(1 - \cos\eta), \quad t = a_0(\eta - \sin\eta).$$

We see that solutions presented here are tightly connected with the gravitational energy. These solutions have singularities and in connection with this we would like to make the following remark.

From the theory of time presented here it follows directly that there is **matter outside the time** . For example, the gravity field and the electromagnetic field exterior to time are described by the equations

$$R_{ij} - \frac{1}{2}g_{ij}R = g_{ij}F^2 - F_{ik}F_{jl}g^{kl},$$

$$\nabla_i F^{ij} = 0, \quad F^{ij} = F_{kl}g^{ik}g^{jl}, \quad F_{ij} = \partial_i A_j - \partial_j A_i,$$

where $F^2 = \frac{1}{4}F_{ij}F^{ij}$. In this context it is very important to understand the nature of the emergence of time.

To conclude this section, it is useful to show the solution for the case when energy-momentum tensor is equal to zero and physical manifold is flat, i.e., the Riemann tensor of the metric (1) is equal to zero, $R_{ijl}{}^{k} = 0$. In such case physical manifold can be considered as four-plane and hence $g_{ij} = \delta_{ij}$. Under such conditions equations (5) has the following regular everywhere solution $f(u) = a_i x^i$, where a_1, a_2, a_3, a_4 are constants constrained by the algebraic equation $a_1^2 + a_2^2 + a_3^2 + a_4^2 = 1$. Thus, gradient of the temporal field is a constant unit vector with arbitrary direction on the four-plane. If one considers this physical manifold as a background, then any physical consequences should not depend on this arbitrariness. Thus, the described solution depends on three arbitrary parameters and it is not difficult to verify that for this solution $\varepsilon_g = 0$. Selecting one direction from the continuum we transmute considered physical manifold into the Minkowski space-time, which as is well-known, is a good background for the description of the wide spectrum of the physical phenomena.

Conclusion

Let us now sum up the obtained results and focus on some of the problems. It is shown that gravity physics is an internal field theory by nature which does not contain *a priori* elements and can be characterized as follows. In the theory there is no internal reason that could objectively distinguish one arbitrary coordinate system from another. It is unrolled on the smooth four-manifold that does not exist *a priori* but is defined by the physical system itself. Manifold is considered as a basic primary notion in physics and representation about space and time is given on this ground. All notions, definitions and laws are formulated in coordinate independent form, i.e., within the framework of the structure of smooth manifold. Einstein's gravitational potentials are determined by the positive definite symmetrical tensor field (Riemann metric) and temporal field. Thus, the whole approach relies on the concepts of genuine Riemannian geometry. Being a scalar field on the manifold, the temporal field is introduced into the theory in the form that provide, in general, time reversal invariance. The system evolves in its proper time and it is important that we do not have compare intrinsic time with some other times. Central role of the time in the internal field theory is that time determines causal structure of the field theory and first of all it characterizes the internal nature of the gravity field as the simplest closed system. It is established that energy density is the first integral of the closed system of interacting fields. The general covariant definition of the electric and magnetic fields is given and the Maxwell equations for these fields are derived. It is shown how to reveal important internal properties of physical system when coordinates have no physical sense. This means that physical sense of general covariance is recognized and there exists a reparametrization-invariant description (see [9] for an excellent and detailed survey on this subject).

References

[1] A. Einstein, *Z. Math. und Phys.*, 225 (1913) (Mit M. Grossmann); Naturforsch. Gesellhaft, Zürich, Vierteljahrschr, 284 (1913); *Phys. Z.*, 1249 (1913); *Z. Math. und Phys.*, 215 (1913) (Mit M. Grossmann); Sitzungsber. preuss. Acad. Wiss., 844 (1915); Ann. Phys., 769 (1916).

[2] L.D. Landau and E.M. Lifshitz, *The Classical Theory of Fields* (Addison-Wesley and Pergamon, New York, 1971).

[3] S. Weinberg, *Gravitation and Cosmology: Principles and Applications of the General Theory of Relativity* (Wiley, New York, 1972).

[4] S.W. Hawking and G.F.R. Ellis, *The large scale structure of Space-Time* (University Press, Cambridge, 1973).

[5] C.W. Misner, K.S. Thorne, and J.A. Wheeler, *Gravitation* (Freeman, San Francisco, 1973).

[6] B.S. DeWitt, Phys. Rev., 1113 (1967).

[7] E. Alvarez, Rev. Mod. Phys., 561 (1989).

[8] C. Rovelli, Phys. Rev. D, 442 (1991).

[9] C. Isham, *Canonical quantum gravity and the problem of time* (Lectures presented at the NATO Advanced Study Institute, Salamanca, June 1992); gr-qc/9210011.

[10] D. Gromoll, W.Klingenberg, and W.Meyer, *Riemannsche Geometrie im Grossen* (Springer-Verlag, 1968).

[11] S.M. Carrol, G.B. Field and R. Jackiw, Phys. Rev. D, 1231 (1990).

[12] P.A.M. Dirac, *General Theory of Relativity* (A Wiley-Interscience Publication, 1975), p. 35.

In: Horizons in World Physics, Volume 248
Editor: A. Reimer, pp. 31-46 ISBN 1-59454-322-4

Chapter 2

COSMIC TIME PARADOX IN QUANTUM COSMOLOGY

E. I. Guendelman and A. B. Kaganovich
Ben Gerion University, Beer Shiva, Israel

The absence of time in the Wheeler-DeWitt (constraint) equation is the main point leading to the generally accepted conclusion that in the quantum cosmology there is no possibility to describe the evolution of the universe proceeding in the cosmic time (the time usually used in classical cosmology). We show that there exist anisotropic quantum cosmology models where in spite of the constraint, the averaging of some of the Heisenberg equations yields *explicit cosmic time dependence of the expectation values of certain dynamical variables*. The reason is that in the context of the minisuperspace approach to quantum cosmology, anisotropic solutions of the Wheeler-DeWitt equation have a feature similar to that of a free nonrelativistic particle solution. The latter implies that these states can be normalized only in a continuous way and a careful definition of the averaging procedure is needed. This "technical detail" leads to the mentioned above appearance of explicit cosmic time dependence of the expectation values of certain dynamical variables, - the result which can be treated at least, as a particular resolution of the conceptual problem of time in quantum gravity. This idea is realized explicitly in two higher dimensional models: in the first model, the sources of gravity are a negative cosmological constant and dust; in the second model - a negative cosmological constant and coherent dilaton excitations. We find the new quantum effect which consists in "quantum inflationary phase" for some dimensions and simultaneous "quantum deflationary" contraction for the remaining dimensions. The expectation value of the "volume" of the universe remains constant during this quantum "inflation-deflation" process. We briefly mention the works of other authors (A. Kheyfets, W.A. Miller, A.P. Gentle and N.D. George; M. Weinstein and R. Akhoury) which also show a possibility of an explicit cosmic time dependence of the expectation values of some of the cosmological variables.

INTRODUCTION

In the quantization of generally covariant systems, like General Relativity (in any number of dimensions), one has to take into account a fundamental constraint which basically tell us that the total Hamiltonian of the classical system equals zero[1], $H \approx 0$. The sign $\approx$ is used[2] instead of $H = 0$ to emphasize the fact that although H is zero along the classical trajectories, it is still non trivial in a sense that it may have non zero Poisson brackets with other dynamical variables.

In the quantized version of the theory, the physical states have to satisfy the constraint equation $H\Psi = 0$ which means that those states should be time independent. One believes usually that the expectation values of the Heisenberg equations for operator of *any* dynamical variable is always zero just by virtue of the constraint equation, i.e. expectation values of all the dynamical variables must be time independent too. Such a situation is interpreted in the literature as the statement (having almost the power of a theorem) that time disappears from quantum gravity[1,3,4]. In particular, in the context of quantum cosmology, this statement is formulated as the generally accepted conclusion that there is no possibility to describe the evolution of the universe proceeding in the cosmic time.

We will see in this chapter that the above conclusions can be premature. This will be done by presentation of an explicit counter-example where cosmic time keeps its original role in quantum cosmology: the expectation values of certain variables have non trivial dependence on the same cosmic time which enters, for instance, in the classical equations of the Friedmann cosmology.

The reason for the recovery of a non trivial time dependence in quantum cosmology, in spite of the fact that H equals zero when applied on physical states, lies on the two complementary facts: a) The Hamiltonian is still a nontrivial operator since it can have non trivial commutators with some of the observables[5] and therefore the Heisenberg equations are reproduced. This means that in such case there must be non physical states $|N.P.\rangle$ for which $H|N.P.\rangle \neq 0$. b) In our specific model, the physical states Ψ, for which $H|\Psi\rangle = 0$ is satisfied, are found to be non normalizable states. This forces us to consider the "non physical" states $|N.P.\rangle$ in order to define expectation values of relevant operators with the help of a limit process $|N.P.\rangle \to |\Psi\rangle$. Since for the non physical states $H|N.P.\rangle \neq 0$, non trivial time dependence can appear in the expectation values of some of the Heisenberg equations.

We will display this interesting phenomenon of the appearance of nontrivial time dependence in quantum cosmology in the context of a Kaluza-Klein model which allows for anisotropic evolution: expansion of 3 dimensions and contraction of the extra dimensions for a physically attractive scenario. In this model, a negative cosmological constant does not let the total volume of the universe grow, while quantum effects stabilize the volume against collapse.

It is important to point out that the appearance of time we display in our model, *does not* rely on some WKB approximation[6] or on the use of some field variable as time[7], rather, it is the *genuine, original cosmic time* which fulfils its natural duty to be the time parameter of the theory even at the intrinsic quantum level.

Description of the Model and its Classical Dynamics

We start from studying a higher dimensional homogeneous totally anisotropic spatially flat cosmological model

$$ds^2 = -dt^2 + \sum_{l=1}^{D} a_l^2(t)dx_l^2 \tag{1}$$

which is assumed to be toroidally compact: $0 \le x_l \le L_l$. The scalar curvature corresponding to Eq.(1) is given by

$$R = \frac{2}{V}\frac{d^2V}{dt^2} - \frac{D-1}{D}\left(\frac{d(\ln V)}{dt}\right)^2 + \frac{1}{D}\sum_{l<m}\left(\frac{d(\ln a_l)}{dt} - \frac{d(\ln a_m)}{dt}\right)^2 \tag{2}$$

where $V = \prod_{l=1}^{D} a_l \equiv$ "volume" of the universe. We will assume that the only sources of gravity are a negative cosmological constant $\Lambda < 0$ and dust. The gravitational action can be written then as

$$S_{gr} = \frac{1}{\kappa}\int d^D x dt \sqrt{-g}(R - 2\Lambda) = -\frac{1}{\kappa}\prod_{l=1}^{D} L_l \int L dt \tag{3}$$

where $\kappa = 16\pi G$ and up to a total derivative term, the Lagrangian L is given by

$$L = \frac{1}{V}\frac{D-1}{D}\left(\frac{dV}{dt}\right)^2 - \frac{V}{D}\sum_{l<m}\left(\frac{d(\ln a_l)}{dt} - \frac{d\ln(a_m)}{dt}\right)^2 + 2\Lambda V \tag{4}$$

For simplicity, we choose units where $\frac{1}{\kappa}\prod_{i-1}^{D} L_l = 1$ so that

$$S_{gr} = -\int L dt \tag{5}$$

In addition to the equations derived from (4) and (5) we have to impose the constraint saying that the Hamiltonian H is zero (the statement which coincides with the $0-0$ component of Einstein's equations). The presence of dust affects only the constraint equation which reads

$$H = \frac{1}{V}\frac{D-1}{D}\left(\frac{dV}{dt}\right)^2 - \frac{V}{D}\sum_{l<m}\left(\frac{d(\ln a_l)}{dt} - \frac{d(\ln a_m)}{dt}\right)^2 - 2\Lambda V - \mu = 0, \tag{6}$$

where $\mu > 0$ is a constant which has the interpretation of the dust energy density times the volume V of the universe.

The form of the Lagrangian and the constraint may be simplified to a marked degree if one uses the following parametrization for $a_l(t)$

$$a_l(t) = [V(t)]^{1/D} e^{\theta_l(t)}. \tag{7}$$

Since $\sum_{i=1}^{D} \theta_l \equiv 0$ one can exclude $\theta_D = -\sum_{i=1}^{D-1} \theta_i$ and proceed with the D independent variables: V and θ_i, i=1,2,...D-1. Finally, one can see that the Lagrangian and the constraint take the diagonalized and normalized form

$$L = \left(\frac{d\rho}{dt}\right)^2 - \rho^2 \sum_{i=1}^{D-1} \left(\frac{dz^i}{dt}\right)^2 - \omega^2 \rho^2, \tag{8}$$

$$H = \left(\frac{d\rho}{dt}\right)^2 - \rho^2 \sum_{i=1}^{D-1} \left(\frac{dz^i}{dt}\right)^2 + \omega^2 \rho^2 - \mu = 0, \tag{9}$$

where

$$\omega^2 = -\frac{D}{2(D-1)}\Lambda; \quad \rho^2 = \frac{4(D-1)}{D} V \tag{10}$$

and

$$z^i = \frac{1}{1+\sqrt{D}} \sqrt{\frac{D}{2(D-1)}} [\theta_1 + \theta_2 + \ldots + (2+\sqrt{D})\theta_i + \ldots + \theta_{D-1}],$$
$$i = 1, 2, \ldots, D-1$$

Notice that the Lagrangian (8) and the constraint (9) are invariant under $\frac{D(D-1)}{2}$ dimensional symmetry group of translations and rotations of a $D-1$ dimensional Euclidean space. In particular, the translational symmetry $z^i \to z^i + b^i$ with constants b^i, i=1,2,...,D-1, gives rise to D-1 conserved quantities

$$F_i = -2\rho^2 \frac{dz^i}{dt}, \quad i = 1, 2, \ldots, D-1 \tag{11}$$

As we will see in what follows, it is very important that the set of quantities $\frac{dz^i}{dt}$ (which are the linear combinations of $\frac{d\theta^i}{dt}$) measure *anisotropy* of the evolution of the universe. In fact, since $\sum_{i-1}^{D}\theta_i \equiv 0$, all θ_i's can have the same time dependence only if such time dependence is the trivial one, i.e., all of the θ_i's are constants. Therefore, the set of $D-1$ conserved independent quantities F_i measure anisotropy of the evolution of the universe.

It is interesting to note that the constraint may be represented now in the following form

$$H=\left(\frac{d\rho}{dt}\right)^2+U_{eff}^{(class)}(\rho)-\mu=0 \tag{12}$$

where the effective volume dependent (classical) potential appears:

$$U_{eff}^{(class)}(\rho)\equiv-\frac{F^2}{4\rho^2}+\omega^2\rho^2,\quad F^2\equiv\sum_{i=1}^{D-1}F_i^2 \tag{13}$$

All classical solutions exhibit cosmological singularities. This feature, as we will see below, can be avoided in quantum cosmology in the presence of negative cosmological constant and dust.

MINISUPERSPACE QUANTIZATION AND SOLUTIONS OF THE WHEELER-DEWITT EQUATION

To produce a quantum theory from the classical one, we have to postulate canonical commutation relations and in addition, the constraint equation (9) has to be imposed as a condition on the wave function of the universe Ψ. The resulting Wheeler-DeWitt equation $H\Psi=0$ is the fundamental equation governing the quantum cosmology.

It is convenient to rewrite the Lagrangian (8) and the Hamiltonian (9) in a geometrical fashion

$$L=f_{\alpha\beta}(q)\frac{dq^\alpha}{dt}\frac{dq^\beta}{dt}-\omega^2\rho^2, \tag{14}$$

$$H=f_{\alpha\beta}(q)\frac{dq^\alpha}{dt}\frac{dq^\beta}{dt}+\omega^2\rho^2-\mu=\frac{1}{4}f^{\alpha\beta}\pi_\alpha\pi_\beta+\omega^2\rho^2-\mu \tag{15}$$

where

$$q^\alpha=(\rho,z^i),\quad i=1,2,\ldots,D-1;\quad f_{\alpha\beta}(q)=diag(1,-\rho^2,\ldots,-\rho^2) \tag{16}$$

are respectively coordinates and metric of our minisuperspace and π_α are the momenta canonically conjugate to q^α.

The ambiguity due to operator ordering problem can be taken into account[8] by adding the "nonminimal" term ξR_f, where R_f is the scalar curvature corresponding to the metric $f_{\alpha\beta}(q)$ and ξ is an arbitrary real constant depending on the operator ordering used. The resulting Wheeler-DeWitt equation reads[9]

$$\frac{1}{\rho^{D-1}}\frac{\partial}{\partial\rho}\left(\rho^{D-1}\frac{\partial\Psi}{\partial\rho}\right)-\frac{1}{\rho^2}\sum_{i=1}^{D-1}\frac{\partial^2\Psi}{\partial z^{i2}}+\left[\xi\frac{(D-1)(D-2)}{\rho^2}-4\omega^2\rho^2+4\mu\right]\Psi=0, \quad (17)$$

where we have used that the scalar curvature of the minisuperspace metric $f_{\alpha\beta}$ is $R_f=(D-1)(D-2)\rho^{-2}$.

The inner product for sufficiently regular wavefunctions Ψ and Φ is defined in the geometric form

$$(\Psi,\Phi)=\int d^D q\sqrt{|\,det(f_{\alpha\beta})\,|}\Psi^*(q)\Phi(q). \quad (18)$$

To solve Eq.(17), let us note that the Hamiltonian commutes with the generators $-i\frac{\partial}{\partial z^i}$ of the symmetry $z^i\to z^i+b^i$ with $D-1$ arbitrary real numbers b^i. It is therefore possible to take the solutions of (17) as eigenstates of the generators $-i\frac{\partial}{\partial z^i}$

$$\Psi(\rho,z^i)=\frac{1}{(2\pi)^{(D-1)/2}}R(\rho)\exp\left(i\sum_{i=1}^{D-1}F_i z^i\right), \quad (19)$$

where F_i is an eigenvalue of the operator $-i\frac{\partial}{\partial z^i}$, which is the quantum version of the conserved quantity defined by Eq.(11). The function $R(\rho)$ is then determined as the solution of the equation

$$\frac{d^2R}{d\rho^2}+\frac{D-1}{\rho}\frac{dR}{d\rho}+\frac{K}{\rho^2}R-4\omega^2\rho^2R=-4\mu R, \quad (20)$$

where

$$K\equiv F^2+\xi(D-1)(D-2) \quad (21)$$

and F^2 is defined in Eq.(13). It is possible to regard Eq.(20) as a stationary Schrodinger equation in a D-dimensional space (with mass of particle $=1/2$) with an effective potential

$$U_{eff}^{(quant)}(\rho)\equiv-\frac{K}{\rho^2}+4\omega^2\rho^2. \quad (22)$$

Depending on whether (i) $K > 0$, (ii) $K = 0$ or (iii) $K < 0$ the effective potential has an attractive core, is exactly a harmonic potential or has a repulsive core correspondingly. Then 4μ in Eq.(20) plays the role of the "energy" eigenvalue. Since the region $\rho \to \infty$ is classically forbidden, a physically acceptable solution of Eq(20) has to vanish in this limit. This leads to the quantization of the eigenvalues of the linear operator corresponding to Eq.(20). However, in our situation, it is not very appealing to quantize the energy of dust μ because this is not a dynamical variable. In contrast to this, the conserved quantity F^2, which is a measure of the anisotropy of the cosmological evolution, can be regarded as the quantized dynamical variable. Using the fact that F^2 appears linearly in Eq.(20), for a *given* μ, we select the appropriate values of F^2 such that the operator corresponding to the l.h.s. of Eq.(20) has as its eigenvalue the number 4μ. Therefore we have different values of F^2 specifying uniquely the eigenfunctions.

The solution of Eq.(20) is given by

$$R = R_n(\rho) = N_n \rho^{2s_n} e^{-|\omega|\rho^2} \Phi\left(-n, \frac{\mu}{|\omega|} - 2n, 2|\omega|\rho^2\right), \tag{23}$$

where N_n is a normalization factor, Φ is a confluent hypergeometric function, n is a non-negative integer and[1]

$$s_n = \frac{1}{4}\left[-(D-2) + \sqrt{(D-2)^2 - 4K_n}\,\right] = \frac{1}{2}\left(\frac{\mu}{|\omega|} - 2n - \frac{D}{2}\right). \tag{24}$$

The corresponding quantized values of the length of the vector F_i are

$$F^2 = F_n^2 = 4\left\{(D-2)\left[\frac{1}{4}(D-2) - \xi(D-1)\right] - \left[\frac{\mu}{|\omega|} - 1 - 2n\right]^2\right\}. \tag{25}$$

and K_n in (24) is determined by F_n^2 via Eq.(21). Notice that the direction of the $(D-1)$-dimensional anisotropy vector F_i remains arbitrary.

Avoidance of the cosmological singularity in the context of quantum cosmology can be defined as a statement that the amplitude $\Psi \to 0$ as $\rho \propto \sqrt{V} \to 0$. In our case, this is possible if $s_n > 0$, i.e.

$$\mu > \left(2n + \frac{D}{2}\right)|\omega|. \tag{26}$$

[1] We do not consider another possible solution with $s_n = \frac{1}{4}\left[-(D-2) - \sqrt{(D-2)^2 - 4K_n}\,\right]$ because it violates several physical restrictions.

We see that the presence of enough amount of dust is a necessarily condition for the avoidance of the cosmological singularity. It follows from Eq.(24) that condition $s_n > 0$ implies also that $K_n < 0$, that is $F_n^2 < -\xi(D-1)(D-2)$. Assuming $D > 2$, this can be achieved for some $F_n^2 > 0$ only if $\xi < 0$. Notice that $K_n < 0$ means that the quantum effective potential (22) has a repulsive core.

THE DYNAMICS OF QUANTUM COSMOLOGY PROCEEDING IN COSMIC TIME

Heisenberg Picture

Before studying the problem of defining the role of time in quantum cosmology processes in the framework of our model, it is important to understand some aspects of the classical dynamics. This is of importance since the quantum behavior has to satisfy, in some way, the correspondence principle. In particular, the time dependence of z^i determined by Eq.(11)

$$\frac{dz^i}{dt} = -\frac{F_i}{2\rho^2}, \quad i = 1, 2, ..., D-1 \tag{27}$$

can be obtained not only from the Euler-Lagrange equations but also from the Hamiltonian formalism, i.e.

$$\frac{dz^i}{dt} = \{z^i, H\} = -\frac{F_i}{2\rho^2}, \quad \frac{d\pi_{z^i}}{dt} = -\frac{d}{dt}\left(2\rho^2 \frac{dz^i}{dt}\right) = \{\pi_{z^i}, H\} = 0 \tag{28}$$

where π_{z^i} is the momenta canonically conjugated to z^i. Note that *although* $H = 0$ *along the classical trajectories,* H *has nontrivial Poisson brackets with* z^i (see the related discussion by Kuchar[5]). This allows a nontrivial (cosmic) time dependence through Eqs.(28), which provides *information not contained in the constraint equation (9).*

We will now see that a similar situation appears in the quantum version of the theory, where so far the only thing we have cared about has been the Wheeler-DeWitt (quantum constraint) equation (17). Normally, when quantizing a theory without constraints, the Poisson bracket $\{A, B\}$ is replaced by $i[A, B]$, where $[A, B]$ is the quantum mechanical commutator of A and B. For any operator Q which does not depend explicitly on time, the Heisenberg equations hold

$$\frac{dQ}{dt} = i[H, Q]. \tag{29}$$

For our purpose, it is convenient to work in the Heisenberg picture, where operators satisfy Eq.(29), but states are taken to be time independent, that is

$$\frac{\partial\Psi}{\partial t}=0. \tag{30}$$

We are interested in a subspace of this space of functions, the so-called "physical subspace", in which the constraint equation (17) holds. The consistency requirement that the constraint be preserved in time, is here trivially satisfied because the constraint function H coincides with the Hamiltonian. Solution of Eq.(17) has been presented in the previous section, and it is now our purpose to look at the consequences of the Heisenberg equations that determine *the evolution of operators defined in the whole space of functions* (and not just states satisfying the constraint $H\Psi=0$).

The operator equations that govern the cosmic time dependence of z^i ($i=1,2,...,D-1$) are

$$\frac{dz^i}{dt}=i[H,z^i]=\frac{i}{4\rho^2}\sum_{k=1}^{D-1}\left[\frac{\partial^2}{\partial z^{k2}},z^i\right]=\frac{i}{2\rho^2}\frac{\partial}{\partial z^i}. \tag{31}$$

$$\frac{d^2z^i}{dt^2}=-\left(\frac{D-4}{\rho^4}+\frac{2}{\rho^3}\frac{\partial}{\partial\rho}\right)\frac{\partial}{\partial z^i}. \tag{32}$$

Quantum Mechanical Averaging

We are going now to evaluate the averages of $\frac{dz^i}{dt}$ and other dynamical variables. However, one should be careful on how averages are defined for wavefunctions (19) since these are normalizable in a continuous way only.

To see that naive manipulations in this case can lead to wrong results, let us consider Eq.(29) together with the subsidiary condition $H\Psi=0$. If we assume H to be an Hermitian operator, we would get (using the inner product (18)) that $\frac{d}{dt}\langle\Psi|Q|\Psi\rangle=\left\langle\Psi|\frac{dQ}{dt}|\Psi\right\rangle=\langle\Psi|i[H,Q]|\Psi\rangle=0$, from where it seems to follow that all averages are time independent. However, the above manipulations are not correct, since the assumption that H is an Hermitian operator holds only if the functions considered are not pathological. In fact, the latter is not always true since if we consider, for example, the case $Q=z^i$, the state $z^i\,|\Psi\rangle$ (that enters in the expression $\langle\Psi|i[H,Q]|\Psi\rangle$), where $|\Psi\rangle$ is given by (19), is very singular in the limit $|z^i|\to\infty$. So, the conclusion about the time independence of the expectation values of *all* canonical operators $Q(q,\pi)$, i.e. that $\frac{d}{dt}\langle\Psi|Q|\Psi\rangle=0$ is always true, has been based on a wrong assumption.

Notice that the z^i-dependence of the wavefunction (19) (resulting from the symmetry $z^i \to z^i + b^i$) resembles other more familiar example in physics, namely the momentum eigenstate $|\vec{p}\rangle$ of a free nonrelativistic particle governed by the Hamiltonian $H = \frac{\hat{\vec{p}}^2}{2m}$. We have in the Heisenberg picture $\frac{d}{dt}\langle \vec{p} | \vec{x} | \vec{p} \rangle = \left\langle \vec{p} | \frac{d}{dt}\vec{x} | \vec{p} \right\rangle = i\langle \vec{p} | [H, \vec{x}] | \vec{p} \rangle$ $= i\langle \vec{p} | (H\vec{x} - \vec{x}H) | \vec{p} \rangle$. Taking into account that $H|\vec{p}\rangle = \frac{\vec{p}^2}{2m}$ and *assuming* H is hermition, we get $\frac{d}{dt}\langle \vec{p} | \vec{x} | \vec{p} \rangle ? = ?i(\frac{\vec{p}^2}{2m} - \frac{\vec{p}^2}{2m})\langle \vec{p} | \vec{x} | \vec{p} \rangle = 0$. The dubious step is actually based on a wrong assumption, when we have taken for granted the hermiticity of H. The condition for the hermiticity of H fails in this case because of the singular behavior of the state $\vec{x}|\vec{p}\rangle$ at large distance and this is the reason why the result is wrong. In fact, we know that the right answer is obtained by replacing $i[H, \vec{x}]$ by $\hat{\vec{p}}/m$, that leads to the conclusion $(\langle \vec{p} | \vec{p} \rangle)^{-1} \frac{d}{dt}\langle \vec{p} | \vec{x} | \vec{p} \rangle = \vec{p}/m$. Although the last argument leads to the right result, the method of calculation is not totally satisfactory, since it involves carelessly manipulating infinite factors like $\langle \vec{p} | \vec{p} \rangle$, etc. The same conclusion can however be obtained more rigorously by defining

$$\frac{1}{\langle \vec{p} | \vec{p} \rangle} \frac{d}{dt}\langle \vec{p} | \vec{x} | \vec{p} \rangle \equiv \lim_{|\varphi\rangle \to |\vec{p}\rangle} \lim_{|\chi\rangle \to |\vec{p}\rangle} \left[\frac{1}{\langle \varphi | \chi \rangle} \frac{d}{dt} \langle \varphi | \vec{x} | \chi \rangle \right] \tag{33}$$

provided $|\varphi\rangle$ and $|\chi\rangle$ are states in the Heisenberg representation which have good enough large distance behavior (e.g. wave packets) so that the integrals appearing in the numerator and denominator in the r.h.s. of (33) are well defined (it means that the states $|\varphi\rangle$ and $|\chi\rangle$ are not eigenstates of the operators $\hat{\vec{p}}$ and H).

Proceeding in the same fashion in our quantum cosmology model, we define in general

$$\langle Q \rangle \equiv \lim_{|\varphi\rangle \to |\Psi\rangle} \lim_{|\chi\rangle \to |\Psi\rangle} \left[\frac{1}{\langle \varphi | \chi \rangle} \langle \varphi | Q | \chi \rangle \right], \tag{34}$$

where $|\Psi\rangle = |F_i, n\rangle$ is the eigenstate of the operator $-i\frac{\partial}{\partial z^i}$ with the eigenvalue F_i and the corresponding wavefunction $\langle \rho, z^i | \Psi \rangle = \Psi(\rho, z^i)$ is the solution (19), (23) of the constraint $H\Psi = 0$; $|\varphi\rangle$ and $|\chi\rangle$ are states in the Heisenberg representation (and therefore static) which have good enough behavior as $|z^i| \to \infty$, so that the integrals appearing in the numerator and denominator of (34) are well defined. This necessarily implies that these states do not satisfy the constraint, i.e. $H\varphi \neq 0$, $H\chi \neq 0$ although for each point (ρ, z^i), the wavefunctions $\langle \rho, z^i | \varphi \rangle$ and $\langle \rho, z^i | \chi \rangle$ approach the wavefunction $\Psi(\rho, z^i)$ in the limit process appearing in (34).

In particular, when $Q = \frac{dz^i}{dt}$, we find

$$\left\langle \frac{dz^i}{dt} \right\rangle = \frac{d\langle z^i \rangle}{dt} \equiv \lim_{|\varphi\rangle \to |\Psi\rangle} \lim_{|\chi\rangle \to |\Psi\rangle} \left[\frac{1}{\langle \varphi | \chi \rangle} \left\langle \varphi | \frac{dz^i}{dt} | \chi \right\rangle \right], where |\Psi\rangle = |F_i, n\rangle. \tag{35}$$

Using the definition (35) to evaluate the average of (31) in a state of the form given by Eq.(19), we get the result

$$\frac{d\langle z^i \rangle}{dt} = -\left\langle \frac{1}{2\rho^2} \right\rangle_n F_i, \tag{36}$$

which corresponds exactly with the classical result (27).

In the same way one can proceed to evaluate other averages. In particular, this way of proceeding gives for functions of ρ alone, $Q(\rho)$:

$$\langle Q(\rho) \rangle_n = \frac{\int Q(\rho) R_n^2(\rho) \rho^{D-1} d\rho}{\int R_n^2(\rho) \rho^{D-1} d\rho}, \tag{37}$$

where $R_n(\rho)$ is determined by Eq.(23). The convergence of $\frac{d\langle z^i \rangle}{dt}$ implies the convergence of $\left\langle \frac{1}{\rho^2} \right\rangle$. One can show that $\left\langle \frac{1}{\rho^2} \right\rangle_n$ is finite provided $\mu > (2n+1)|\omega|$ and

$$\left\langle \frac{1}{\rho^2} \right\rangle_n = \frac{2\omega^2}{\mu - (2n+1)|\omega|}. \tag{38}$$

It turns out that the average of $\frac{d^2 z^i}{dt^2}$ in a state given by Eqs.(19) and (23) is identically zero provided

$$\mu > 2(n+1)|\omega| \tag{39}$$

One can show that the restriction (39) on the amount of dust is a necessary condition to provide the consistency of the main three items of our above analyzes: the constraint equation $H\Psi = 0$, the Heisenberg equations and definition of averages.

Recall that the condition (26) is needed if we require that the universe have a zero probability amplitude of having zero volume. Comparing (39) with the condition (26), we conclude that for $D \geq 4$, the quantum cosmology problem under consideration has a satisfactory solution if the condition (26) is satisfied. We will assume it in what follows.

It is very important also that the condition (26) is a stronger restriction on the amount of dust than is actually needed to provide that *the average of the "volume" of the universe* $\langle V\rangle_n = \frac{D}{4(D-1)}\langle \rho^2\rangle_n$ is finite and it turns out to be *time independent*:

$$\langle V\rangle_n = \frac{D}{4(D-1)}\frac{\int R_n^2(\rho)\rho^{D+1}d\rho}{\int R_n^2(\rho)\rho^{D-1}d\rho} = \frac{D}{16(D-1)\omega^2}\frac{\mu^2-(2n+1)^2\omega^2}{\mu-(2n-1)|\omega|} \tag{40}$$

In particular, for the ground state, $n = 0$, $\langle V\rangle_{n=0} = \frac{D}{16(D-1)\omega^2}(\mu - |\omega|)$. The integrals appearing in (40) when they converge of course give a positive result as it must be from the definition of $\langle V\rangle_n$.

The Cosmic Time Dependence of the Expectation Values of the Cosmological Quantities: Some General Results

To see more clear the above results, let us now represent them in terms of usual cosmological quantities. In the totally anisotropic model (1), the corresponding classical variables are the "scale factors" $a_l(t)$, (l=1,2,...,D) which have been parametrized by means of Eq.(7) in terms of the "volume" of the universe $V(t)$ and D functions $\theta_l(t)$ (l=1,2,...,D). Remind that due to the identity $\sum_{i-1}^{D}\theta_l \equiv 0$, only $D-1$ of the θ_l's are independent . Relations between variables V and θ_i $(i = 1,2,...,D-1)$ and those ρ and z^i are given by Eqs.(10) and (11).

In quantum cosmology, the average of z^i, even with the improved definition (34), does not exist and therefore (34) has nothing to say on whether $\langle z_i\rangle$ is time dependent or time independent. In contrast to this, the average of $\left\langle \frac{dz^i}{dt}\right\rangle$ is well defined and in the presence of an anisotropy ($F_i \neq 0$), it is a nonzero finite constant determined by Eqs.(36) and (38). Then we find that the time evolution of $\langle z_i\rangle$ is of the form

$\langle z_i \rangle = \left\langle \frac{dz^i}{dt} \right\rangle t + c^i = \left(-\left\langle \frac{1}{2\rho^2} \right\rangle_n F_i \right) t + c^i$, where c^i are undetermined constants. This yields to the following cosmic time dependence of the θ_i variables

$$\langle \theta_i \rangle = \alpha_i t + \gamma_i, \tag{41}$$

where γ_i are integration constants and for $i = 1, 2, ..., D-1$

$$\alpha_i = \frac{\sqrt{D-1}}{D(\sqrt{D}+1)} \frac{2\omega^2}{\mu - (2n+1)|\omega|} \left[(D + \sqrt{D} - 1)F_i - \sum_{j \neq i} F_j \right] \tag{42}$$

From the identity $\sum_{l=1}^{D} \theta_l \equiv 0$ we have for $\langle \theta_D \rangle$

$$\langle \theta_D \rangle = \alpha_D t + \gamma_D \quad where \quad \alpha_D = -\sum_{i=1}^{D-1} \alpha_i, \quad \gamma_D = -\sum_{i=1}^{D-1} \gamma_i \tag{43}$$

Besides, in the classical cosmology one can define D expansion parameters $H_l \equiv \frac{1}{a_l(t)} \frac{da_l(t)}{dt}$. In the quantum version of the theory, we have of course to define the ordering of the operators a_l and $\frac{da_l}{dt}$. For example, $a_l^{-1} \frac{da_l}{dt}$ or $a_l^{-1/2} \frac{da_l}{dt} a_l^{-1/2}$, etc. give quantum mechanically distinct definitions for H_l. We will choose a definition

$$H_l \equiv \frac{d}{dt}(\ln a_l), \qquad l = 1, ...D \tag{44}$$

Using this definition, the parametrization (7) and results of the previous subsection one can evaluate *the expectation values of the expansion parameters* $\langle H_l \rangle$ *which turn out to be constants*:

$$\begin{aligned} \langle H_l \rangle &= \left(\frac{D}{4(D-1)} \right)^{1/D} \left(\left\langle \frac{d}{dt} (\ln \rho^{2/D}) \right\rangle + \left\langle \frac{d\theta_l}{dt} \right\rangle \right) \\ &= \left(\frac{D}{4(D-1)} \right)^{1/D} \left\langle \frac{d\theta_l}{dt} \right\rangle = \left(\frac{D}{4(D-1)} \right)^{1/D} \alpha_l, \end{aligned} \tag{45}$$

where α_l, (l=1,...D) are determined in (42) and (43).

Let us notice that Eqs.(7) and (44) together with the identity $\sum_{l-1}^{D}\theta_l \equiv 0$ imply that

$$\sum_{l=1}^{D} H_l = \frac{d}{dt}(\ln V). \tag{46}$$

Due to Eqs.(45) and (43), averaging the l.h.s. of Eq.(46) yields the result that *the sum of the averages of the expansion parameters equals zero*. The same result is obtained when averaging the r.h.s. of Eq.(46) by means of the definition (37) (compare for example with Eq.(40)). This shows that the definition (44) of the expansion parameters is consistent with the quantum stabilization of the volume of the universe, $\langle V \rangle = const$, Eq.(40).

From (44) and (45), we see that the time behavior of $\langle \ln a_l \rangle$ is given by

$$\langle \ln a_l \rangle = \left(\frac{D}{4(D-1)} \right)^{1/D} \alpha_l t + \tilde{\gamma}_l, \quad l = 1,2,...,D, \tag{47}$$

where $\tilde{\gamma}_l$ are arbitrary integration constants.

Inflation-Compactification as a Quantum Effect

We will now see that the results of Sec.5 allow to realize a dynamical explanation of the asymmetry in the sizes of extra and ordinary dimensions in the context of quantum cosmology.

At the classical level, there is no difference on whether we use a_l or $\ln a_l$ as our variables. If Eq.(47) were to hold classically, we could conclude that some dimensions exponentially expand and others exponentially contract, depending on the sign of α_l, as given by (42) and (43). In our case, the behavior of the universe is intrinsically quantum mechanical and we will refer to a "*quantum inflationary phase*" for a given dimension l if the expectation value of the expansion parameter $\langle H_l \rangle = const > 0$. Likewise we will refer to a "*quantum deflationary phase*" for a given dimension l if the expectation value of the expansion parameter $\langle H_l \rangle = const < 0$.

A case of particular interest is when the expectation values of the expansion parameters of three of the dimensions are identical and at the same time the expectation values of the expansion parameters of the remaining $D-3$ spatial dimensions are also identical. In such a case $\alpha_1 = \alpha_2 = \alpha_3 \equiv \alpha$ and $\alpha_4 = \alpha_5 = ... = \alpha_D \equiv \tilde{\alpha}$. Then it follows from the identity $\sum_{l-1}^{D}\alpha_l = 0$ that $\tilde{\alpha} = -3\alpha/(D-3)$ and we get by using Eq.(42):

$$\alpha = \alpha(n) = \pm \frac{2\omega^2}{D[\mu - (2n+1)|\omega|]} \sqrt{\frac{(D-3)(D-1)}{3}} \, |F|_n, \tag{48}$$

$$\tilde{\alpha} = \tilde{\alpha}(n) = \mp \frac{2\omega^2}{D[\mu - (2n+1)|\omega|]} \sqrt{\frac{3(D-1)}{D-3}} \, |F|_n, \tag{49}$$

where $|F|_n \equiv \sqrt{F_n^2}$ and F_n^2 is determined by Eq.(25).

Invoking our definitions of quantum inflationary phase and of quantum deflationary phase, we observe that one set of dimensions is in a quantum inflationary phase and simultaneously another set of dimensions is in quantum deflationary phase. This situation is described by the following equations:

$$\begin{aligned} \langle H_i \rangle &= \alpha(n) \quad for \quad i = 1,2,3; \\ \langle H_j \rangle &= \tilde{\alpha}(n) \quad for \quad j = 4,\ldots,D. \end{aligned} \tag{50}$$

According to Eqs.(48) and (49), choosing the three dimensional subspace to be expanding one, we get the simultaneous contraction of the extra dimensions. During this quantum "inflation-compactification" process, the expectation value of the "volume" of the universe, V, remains constant determined by Eq.(40).

DISCUSSION

The minisuperspace model of quantum cosmology we discussed here shows that the widespread belief that cosmic time, which one uses in classical cosmology, disappears in quantum cosmology altogether, seems to be not always right. In this model we have seen that quantum mechanical averages of certain cosmological quantities can explicitly depend on the same cosmic time which was used in the corresponding classical cosmological model. The non-normalizability of the solutions of the Wheeler-DeWitt equation $H|\Psi\rangle = 0$ requires a certain regularization procedure in the evaluation of the matrix elements (see Eq.(34)). In the intermediate stage of this procedure, one should consider normalizable states which do not satisfy the constraint. This regularization procedure plays a key role for obtaining explicit cosmic time dependence of the expectation values of certain cosmological variables. It is interesting that there is a correspondence between the quantum cosmology results for the cosmic time dependence of the anisotropy, Eq.(36), on the one hand, and results of the classical calculations, Eq.(27).

It has been found that quantum effects stabilize the volume of the universe, so that there can be avoidance of the cosmological singularity. The stabilization of volume is consistent with a new quantum effect: existence of a quantum inflationary phase for some dimensions and simultaneous quantum deflationary phase for the remaining dimensions. This effect can be responsible for a visible asymmetry between ordinary and extra dimensions. One can show[10] that the above results also follow if instead of dust we introduce a massive scalar field whose homogeneous degrees of freedom are described quantum mechanically.

A somewhat related approach which also gives cosmic time dependence for averages of certain dynamical variables in the presence of anisotropy was developed by Kheyfets and Miller in Refs.[11],[12]. For the generalization of this approach see Ref.[13].

The observation that the source of the problem of time and its solution are related to the fact that the condition $H\,|\,\Psi\rangle = 0$ implies in general the non-normalizability of the state $|\,\Psi\rangle$ was made also by Weinstein and Akhoury[14]. These authors study the quantum cosmology model associated to inflationary scenarios and are able to obtain explicit cosmic time dependence by the consideration of states that do not satisfy the constraint $H\,|\,\Psi\rangle = 0$.

REFERENCES

[1] K. Kuchar, in Proceedings of the 4th Canadian Conference on General Relativity and Relativistic Astrophysics, ed. G. Kunstatter, D.E. Vincent and J.G. Williams (World Scientific, Singapore, 1992).
[2] P.A.M. Dirac, Lectures on Quantum Mechanics, (Belfer Graduate School of Science, Yeshiva University, New York, 1964)
[3] Lee Smolin, in: "Time and the Instant", Robin Durie (ed.) Manchester, Clinamen Press, 2000; gr-qc/0104097.
[4] J.J. Halliwell, in: "Cambridge 2002, The future of theoretical physics and cosmology", 675-692; gr-qc/0208018.
[5] K. Kuchar, "Canonical quantum gravity", gr-qc/9304012.
[6] Vilenkin, Phys. Rev. D, 1116 (1989).
[7] Hosoya and M. Morikawa, Phys. Rev. D, 1123 (1989).
[8] K. Kuchar, J. Math. Phys., 2122 (1983).
[9] E.I. Guendelman and A.B. Kaganovich, Int. J. Mod. Phys. D, 221 (1993).
[10] E.I. Guendelman and A.B. Kaganovich, Mod. Phys. Lett. A, 1141 (1994).
[11] Kheyfets and W.A. Miller, Phys. Rev. D, 493 (1995).
[12] Kheyfets and W.A. Miller, Int. J. Mod. Phys. A, 4125 (2000).
[13] A.P. Gentle, N.D. George, A. Kheyfets and W.A. Miller, gr-qc/0302044; gr-qc/0302051
[14] M. Weinstein and R. Akhoury, hep-th/0311189; 0312249.

In: Horizons in World Physics, Volume 248
Editor: A. Reimer, pp. 47-73
ISBN 1-59454-322-4

Chapter 3

COVER SCHEMES, FRAME-VALUED SETS AND THEIR POTENTIAL USES IN SPACETIME PHYSICS

John L. Bell[1]
Department of Philosophy, University of Western Ontario, London, Ontario, Canada

ABSTRACT

In the present paper, the concept of a covering is presented and developed. The relationship between cover schemes, frames (complete Heyting algebras), Kripke models, and frame-valued set theory is discussed. Finally cover schemes and frame-valued set theory are applied in the context of Markopoulou's account of discrete spacetime as sets "evolving" over a causal set. We observe that Markopoulou's proposal may be effectively realized by working within an appropriate frame-valued model of set theory. We go on to show that, within this framework, cover schemes may be used to force certain conditions to prevail in the associated models: for example, rendering the universe timeless, obliterating a given event or forcing it to become the universe's "beginning".

PREAMBLE

The concept of *Grothendieck (pre)topology* or *covering* issued from the efforts of algebraic geometers to study "sheaf-like" objects defined on categories more general than the lattice of open sets of a topological space (see, e.g. [4]). A *Grothendieck pretopology* on a category C with pullbacks is defined by specifying, for each object U of C, a set $P(U)$ of arrows to U called *covering families* satisfying appropriate category theoretic versions of the corresponding conditions for a family A of sets to cover a set U, namely: (i) $\{U\}$ covers U, (ii) if A covers U and $V \subseteq U$, then $\mathrm{A} \mid V = \{A \cap V: A \in \mathrm{A}\}$ covers V, and (iii) if A covers U and, for each $A \in \mathrm{A}$, B_A covers A, then $\bigcup_{A \in \mathrm{A}} \mathrm{B}_A$ covers U. In the present paper the covering concept—here called a *cover scheme*—is presented and developed in the simple case when the underlying category is a preordered set. The relationship between cover schemes, frames (complete Heyting algebras), Kripke models, and frame-valued set theory is discussed. Finally cover schemes and frame-valued set theory are applied in the context of Markopoulou's [5] account of discrete spacetime as sets "evolving" over a causal set.

COVER SCHEMES ON PREORDERED SETS

A *preordered set* is a set equipped with a reflexive transitive relation $\leq$. Let $(P, \leq)$ be a fixed but arbitrary preordered set: we shall use letters p, q, r, s, t to denote elements of p. We write $p \cong q$ for $(p \leq q \;\&\; q \leq p)$. A *meet* for a subset S of P is an element p of P such that $\forall q[\forall s \in S(q \leq s) \leftrightarrow q \leq p]$: if p and p' are both meets for S, then $p \cong p'$. If the empty subset $\varnothing$ has a meet, any such meet m is necessarily a *largest* or *top* element of P, that is, satisfies $p \leq m$ for all p. We use the symbol 1 to denote a top element of P. A meet of a finite subset $\{p_1, \ldots, p_n\}$ of P will be denoted by $p_1 \wedge \ldots \wedge p_n$. P is a *lower semilattice* if each nonempty finite subset of P has a meet. A subset S of P is said to be a *sharpening of,* or to *sharpen,* a subset T of P, written $S \prec T$, if $\forall s \in S \exists t \in T(s \leq t)$. A *sieve* in P is a subset S such that $p \in S$ and $q \leq p$ implies $q \in S$. Each subset S of P generates a sieve $\overline{S}$ given by $\overline{S} = \{p : \exists s \in S(p \leq s)\}$.

A *cover scheme* on P is a map $\mathbf{C}$ assigning to each $p \in P$ a family $\mathbf{C}(p)$ of subsets of $p\!\downarrow = \{q\colon q \leq p\}$, called ($\mathbf{C}$-)*covers of* p, such that, if $q \leq p$, any cover of p can be sharpened to a cover of q, i.e.,

$$S \in \mathbf{C}(p) \;\&\; q \leq p \rightarrow \exists T \in \mathbf{C}(q)[\forall t \in T \exists s \in S(t \leq s)]. \qquad \text{(Cov)}$$

If P is a lower semilattice, a *coverage* (see [3]) on P is a map $\mathbf{C}$ as above, satisfying, in place of **(Cov)**, the condition

$$S \in \mathbf{C}(p) \;\&\; q \leq p \rightarrow S \wedge q = \{s \wedge q : s \in S\} \in \mathbf{C}(p).$$

A cover scheme $\mathbf{C}$ is said to be *normal* if every member of every $\mathbf{C}(p)$ is a sieve and whenever $S \in \mathbf{C}(p)$ and T is a sieve such that $S \subseteq T \subseteq p\!\downarrow$, we have $T \in \mathbf{C}(p)$. Any cover scheme $\mathbf{C}$ on P induces a normal cover scheme $\overline{\mathbf{C}}$ (called its *normalization*) defined by

$$\overline{\mathbf{C}}(p) = \{X \subseteq p\!\downarrow : X \text{ is a sieve } \& \;\exists S \in \mathbf{C}(p).\; S \subseteq X\}.$$

Notice that a normal cover scheme on a lower semilattice is always a coverage. For if $\mathbf{C}$ is such, then for $S \in \mathbf{C}(p)$ and $q \leq p$, any sharpening of S to a member of $\mathbf{C}(q)$ is easily seen to be included in $S \wedge q$, so that the latter is also in $\mathbf{C}(q)$.

Write $\mathrm{Cov}(P)$ for the set of all cover schemes on P. There is a natural partial ordering $\lhd$ on $\mathrm{Cov}(P)$ defined by

[1] jbell@uwo.ca

$$\mathbf{C} \lhd \mathbf{D} \leftrightarrow \forall p\ \mathbf{C}(p) \subseteq \mathbf{D}(p)\ .$$

With this ordering Cov(P) is a complete lattice in which the join $\underset{i \in I}{2}\ \mathbf{C}_i$ of any family $\{\mathbf{C}_i : i \in I\}$ is given by

$$(\underset{i \in I}{2}\ \mathbf{C}_i)(p) = \bigcup_{i \in I} \mathbf{C}_i(p).$$

There is also a natural *composition* ★ defined on Cov(P). For $\mathbf{C}, \mathbf{D} \in$ Cov(P), $\mathbf{D}$★$\mathbf{C}$ is defined by decreeing that ($\mathbf{D}$★$\mathbf{C}$)(p) is to consist of all subsets of $p\!\downarrow$ of the form $\bigcup_{s \in S} T_s$, where $S \in \mathbf{C}(p)$ and, for each $s \in S$, $T_s \in \mathbf{D}(s)$. That $\mathbf{D}$★$\mathbf{C}$ is a cover scheme on P may be verified (using the axiom of choice) as follows. Given $S \in \mathbf{C}(p)$, $\bigcup_{s \in S} T_s \in$ ($\mathbf{D}$★$\mathbf{C}$)(p) and $q \le p$, there is $U \in \mathbf{C}(q)$ with $U \prec S$, so for each $u \in U$ there is $s(u) \in S$ for which $u \le s(u)$. Then $T_{s(u)} \in \mathbf{D}(s(u))$ and we can choose $V_u \in \mathbf{D}(u)$ so that $V_u \prec T_{s(u)}$. Clearly $\bigcup_{u \in U} V_u \in$ ($\mathbf{D}$★$\mathbf{C}$)(q) and, since $V_u \prec T_{s(u)}$ for all $u \in U$, it follows immediately that $\bigcup_{u \in U} V_u \prec \bigcup_{s \in S} T_s$.

It is not hard to verify that ★ is associative and that with this operation Cov(P) is actually a *quantale* (see, e.g., [6]) that is, for any $\mathbf{D}$, $\{\mathbf{C}_i : i \in I\}$ in Cov(P),

$$\mathbf{D}\bigstar \underset{i \in I}{2}\ \mathbf{C}_i = \underset{i \in I}{2}\,(\mathbf{D}\bigstar\mathbf{C}_i) \quad (\underset{i \in I}{2}\ \mathbf{C}_i)\bigstar\mathbf{D} = \underset{i \in I}{2}\,(\mathbf{C}_i\bigstar\mathbf{D})\ .$$

Also the element $\mathbf{1} \in$ Cov(P) with $\mathbf{1}(p) = \{p\}$ acts as a quantal unit, since it is readily verified that $\mathbf{1}$★$\mathbf{C}$ = $\mathbf{C}$★$\mathbf{1}$ = $\mathbf{C}$ for all $\mathbf{C} \in$ Cov(P).

In this connection a *Grothendieck pretopology*—which we shall abbreviate simply to *pretopology*—on P may be identified as a cover scheme $\mathbf{C}$ on P satisfying $\mathbf{1}$ Ÿ $\mathbf{C}$ and $\mathbf{C}$★$\mathbf{C}$ Ÿ $\mathbf{C}$, that is, $\{p\} \in \mathbf{C}(p)$ for all $p \in P$ and, if $S \in \mathbf{C}(p)$ and, for each $s \in S$, $T_s \in \mathbf{C}(s)$, then $\bigcup_{s \in S} T_s \in \mathbf{C}(p)$.

We observe that a normal pretopology $\mathbf{C}$ has the additional properties: (i) each $\mathbf{C}(p)$ is a filter of sieves in $p\!\downarrow$, that is, satisfies $S, T \in \mathbf{C}(p) \leftrightarrow S \in \mathbf{C}(p)\ \&\ T \in \mathbf{C}(p)$; (ii) $S \in \mathbf{C}(p)\ \&\ q \le p \rightarrow S \cap q\!\downarrow\ \in \mathbf{C}(q)$. For (ii), we observe that $S \cap q\!\downarrow$, including as it does any sharpening of S to a member of $\mathbf{C}(q)$, is itself a member of $\mathbf{C}(q)$. As for (i), the "→" direction is obvious; conversely, if $S, T \in \mathbf{C}(p)$, then $S \cap T = S \cap \bigcup_{t \in T}(t\downarrow) = \bigcup_{t \in T}(S \cap t\downarrow)$. But from (ii) we have $S \cap t\!\downarrow\ \in \mathbf{C}(t)$ for every $t \in T$, whence $\bigcup_{t \in T}(S \cap t\downarrow) \in \mathbf{C}(p)$, and so $S \cap T \in \mathbf{C}(p)$.

A normal pretopology is also called a *Grothendieck topology*. A normal cover scheme satisfying (i) and (ii) is called a *regular* cover scheme.

Each cover scheme **C** generates a pretopology, and a Grothendieck topology in the following way. First, define $\mathbf{C}^n$ for $n \in \omega$ recursively by $\mathbf{C}^0 = \mathbf{1}$ and $\mathbf{C}^{n+1} = \mathbf{C}\bigstar\mathbf{C}^n$. Now put $\mathbf{G} = 2_{n\in\omega}\ \mathbf{C}^n$. Then **G** is a pretopology, for obviously **1**Ÿ **G,** and

$$\mathrm{G} \bigstar \mathrm{G} = 2_{n\in\omega}\ \mathbf{C}^n \bigstar 2_{m\in\omega}\ \mathbf{C}^m = 2_{n\in\omega}\ 2_{m\in\omega}\ \mathbf{C}^{m+n} = 2_{n\in\omega}\ \mathbf{C}^n = \mathrm{G}.$$

Also **C** Ÿ **G**, and **G** is evidently the Ÿ-least such pretopology. **G** is called the *pretopology generated by* **M.** The normalization $\overline{\mathbf{G}}$ of **G** is then a Grothendieck topology called the *Grothendieck topology generated by* **C.**

Now let **M** be a map assigning to each $p \in P$ a subset $\mathbf{M}(p)$ of subsets of $p{\downarrow}$. Since Cov(P) is a complete lattice, there is a Ÿ-least cover scheme **C** such that $\mathbf{M}(p) \subseteq \mathbf{C}(p)$ for all p. **C** is called the cover scheme *generated by* **M**; the pretopology and Grothendieck topology generated in turn by **C** are said to be *generated by* **M.**

There are several naturally defined cover schemes on P which also happen to be pretopologies. First, each sieve A in P determines two cover schemes $\mathbf{C}_A$ and $\mathbf{C}^A$ defined by

$$S \in \mathbf{C}_A(p) \leftrightarrow p \in A \cup S \qquad\qquad S \in \mathbf{C}^{\mathbf{A}}(p) \leftrightarrow p{\downarrow} \cap A \subseteq S:$$

these are easily shown to be pretopologies. Notice that $\varnothing \in \mathbf{C}_A(p) \leftrightarrow p \in A$ and $\varnothing \in \mathbf{C}^A(p) \leftrightarrow p{\downarrow} \cap A = \varnothing$.

Next, we have the *dense cover scheme* **Den** given by:

$$S \in \mathbf{Den}(p) \leftrightarrow \forall q \leq p \exists s \in S \exists r \leq s(r \leq q): \qquad (*)$$

it is a straightforward exercise to show that this is a pretopology. When S is a sieve, the above condition (*) is easily seen to be equivalent to the familiar condition of *density below p*: that is, $\forall q \leq p \exists s \in S(s \leq q)$.

Note that the following are equivalent for any cover scheme **C:** (a) **C** Ÿ **Den,** (b) $\varnothing \notin \mathbf{C}(p)$ for all p. For since $\varnothing \notin \mathbf{Den}(p)$, (a) clearly implies (b). Conversely, assume (b), and let $S \in \mathbf{C}(p)$. Then for each $q \leq p$ there is $T \in \mathbf{C}(q)$ for which $\forall t \in T\ \exists s \in S\ (t \leq s)$. Since (by (b)) $T \neq \varnothing$, we may choose $t_0 \in T$ and $s_0 \in S$ for which $t_0 \leq s_0$. Since $t_0 \leq q$, and $q \leq p$ was arbitrary, it follows that S satisfies the condition (*) above for membership in $\mathbf{Den}(p)$. This gives (a).

Finally, we have the *Beth cover scheme* **Bet**. This is defined as follows. First we define a *road* from p to be a maximal linearly preordered subset of $p{\downarrow}$: clearly any road from p contains p. Let us call a *rome* over p any subset of $p{\downarrow}$ intersecting every road from p. Now the Beth coverage has $\mathbf{Bet}(p)$ = collection of all romes over p. Let us check first that **Bet** is a cover scheme. Suppose that S is a rome over p and $q \leq p$. We claim that

$$T = \{t \leq q : \exists s \in S(t \leq s)\}$$

is a rome over q. For let Y be any road from q; then, by Zorn's lemma, Y may be extended to a road X from p. We note that since $X \cap q{\downarrow}$ is linearly preordered and includes Y, it must coincide with Y. Since S is a rome over p, there must be an element $s \in S \cap X$. Since also $q \in Y \subseteq X$, we have $s \leq q$ or $q \leq s$. If $s \leq q$, then $s \in X \cap q{\downarrow} = Y$ and $s \in T$, so that $s \in Y \cap T$. If $q \leq s$, then $q \in T$; since $q \in Y$, it follows that $q \in Y \cap T$. So in either case $Y \cap T \neq \varnothing$; therefore T is a rome over q. Since clearly also $T \prec S$, we have shown that **Bet** is a cover scheme.

To show that **Bet** is a pretopology, we observe first that, for any p, $\{p\}$ is a rome over p. Now suppose that we are given: a rome S over p, for each $s \in S$, a rome T_s over s, and a road X from p. Then $s \in X \cap S$ for some s: we claim that $X \cap s{\downarrow}$ is a road from s. For suppose $t \leq s$ is comparable with every member of $X \cap s{\downarrow}$; now since $s \in X$, for each $x \in X$ either $s \leq x$ or $x \leq s$. In the first case $t \leq x$; in the second t is comparable with x by assumption. Hence t is comparable with every member of X, and so $t \in X$. Accordingly $X \cap s{\downarrow}$ is, as claimed, a road from s; as such, it must meet the rome T_s, so X meets $\bigcup_{s \in S} T_s$, and the latter is therefore a rome over p. So **Bet** is indeed a pretopology.

Since clearly $\varnothing \notin$ **Bet**(p) for any p, it follows from what we have noted above that **Bet** Ÿ **Den**, a fact that can also be easily verified directly.

Any preordered set $(P, \leq)$ generates a *free lower semilattice* $\widetilde{P}$ which may be described as follows. The elements of $\widetilde{P}$ are the finite subsets of P; the preordering on $\widetilde{P}$ is the *refinement* relation $\sqsubseteq$, that is, for $F, G \in \widetilde{P}$,

$$F \sqsubseteq G \leftrightarrow \forall q \in G \exists p \in F (p \leq q).$$

The meet operation $\wedge$ in $\widetilde{P}$ is set-theoretic union; the canonical embedding of P into $\widetilde{P}$ is the map $p \mapsto \{p\}$. Notice also that $\varnothing$ is the unique top element of $\widetilde{P}$.

Now, suppose we are given a cover scheme **C** on P. This induces a cover scheme $\tilde{\mathbf{C}}$ on $\widetilde{P}$ defined in the following way. We start by setting $\tilde{\mathbf{C}}(\varnothing) = \{\{\varnothing\}\}$. Now fix a nonempty finite subset F of P, take any nonempty subset $\{p_1, \ldots, p_n\}$ of F and any $S_1 \in \mathbf{C}(p_1), \ldots, S_n \in \mathbf{C}(p_n)$. Define

$$S_1 \bullet \ldots \bullet S_n = \{\{s_1, \ldots, s_n\} \cup F : s_1 \in S_1, \ldots, s_n \in S_n\}.$$

We decree that $\tilde{\mathbf{C}}(F)$ is to consist of all sets of the form $S_1 \bullet \ldots \bullet S_n$, for $S_1 \in \mathbf{C}(p_1), \ldots, S_n \in \mathbf{C}(p_n)$, and all nonempty finite subsets $\{p_1, \ldots, p_n\}$ of F.

Let us check that $\tilde{\mathbf{C}}$ is a cover scheme on $\widetilde{P}$. To begin with, the unique cover $\{\varnothing\}$ of $\varnothing$ is clearly sharpenable to any cover $S_1 \bullet \ldots \bullet S_n$ of any nonempty member of $\widetilde{P}$. Now suppose that $S_1 \bullet \ldots \bullet S_n$ is a $\tilde{\mathbf{C}}$-cover of a nonempty member F of $\widetilde{P}$ and that $G = \{q_1, \ldots, q_m\} \sqsubseteq F$. Then for each $1 \leq i \leq n$ there is $q_i \in G$ for which $q_i \leq p_i$, hence $T_i \in \mathbf{C}(q_i)$ with $T_i \prec S_i$. Clearly $T_1 \bullet \ldots \bullet T_n \in \tilde{\mathbf{C}}(G)$. Also

$T_1 \bullet \ldots \bullet T_n \prec S_1 \bullet \ldots \bullet S_n$. For, given $t_1 \in T_1, \ldots, t_n \in T_n$, then since $T_i \prec S_i$ for each i, there are $s_1 \in S_1, \ldots, s_n \in S_n$ for which $t_1 \le s_1, \ldots, t_n \le s_n$, whence $\{t_1, \ldots, t_n\} \cup G \sqsubseteq \{s_1, \ldots, s_n\} \cup F$. So $\tilde{\mathbf{C}}$ satisfies the conditions of a cover scheme.

The normalization $\bar{\tilde{\mathbf{C}}}$ of $\tilde{\mathbf{C}}$ is then a coverage on $\widetilde{P}$ called the coverage on $\widetilde{P}$ *induced* by **C**.

We next show how cover schemes give rise to complete Heyting algebras, or frames (see, e.g. [3]).

A *Heyting algebra* is a lattice L with top and bottom elements 1, 0 such that, for any elements $x, y \in L$, there is an element—denoted by $x \Rightarrow y$—of L such that, for any $z \in L$,

$$z \le x \Rightarrow y \text{ iff } z \wedge x \le y.$$

Thus $x \Rightarrow y$ is the *largest* element z such that $z \wedge x \le y$. So in particular, if we write $\neg x$ for $x \Rightarrow 0$, then $\neg x$ is the largest element z such that $x \Rightarrow z = 0$: it is called the *pseudocomplement* of x. A *Boolean algebra* is a Heyting algebra in which $\neg\neg x = x$ for all x, or equivalently, in which $x \vee \neg x = 1$ for all x.

If we think of the elements of a (complete) Heyting algebra as "truth values", then 0, 1, $\wedge$, $\vee$, $\neg$, $\Rightarrow$, ı, I represent "true", "false", "and", "or", "not" and "implies", "there exists" and "for all", respectively. The laws satisfied by these operations in a general Heyting algebra correspond to those of *intuitionistic logic.* In Boolean algebras the counterpart of the law of excluded middle also holds.

A basic fact about *complete* Heyting algebras is that the following identity holds in them:

$$x \wedge 2_{i \in I}\, y_i = 2_{i \in I}\,(x \wedge y_i) \qquad (*)$$

And conversely, in any complete lattice satisfying (*), defining the operation $\Rightarrow$ by $x \Rightarrow y =$ ı$\{z\colon z \wedge x \le y\}$ turns it into a Heyting algebra.

In view of this result a complete Heyting algebra is frequently defined to be a complete lattice satisfying (*). A complete Heyting algebra is briefly called a *frame.*

Now we associate a frame with each cover scheme on P. First, we define $\widehat{P}$ to be the set of sieves in P partially ordered by inclusion: $\widehat{P}$ is then a frame—the *completion*[2] of P— in which joins and meets are just set-theoretic unions and intersections, and in which the operations $\Rightarrow$ and $\neg$ are given by

$$I \Rightarrow J = \{p : I \cap p\!\downarrow\, \subseteq J\} \qquad\qquad \neg I = \{p : I \cap p\!\downarrow\, = \varnothing\}.$$

Given a cover scheme **C** on P, a sieve I in P is said to be **C**-*closed* if

---

[2] Writing **Lat** for the category of complete lattices and join preserving homomorphisms, $\widehat{P}$ is in fact the object in **Lat** freely generated by P.

$$\exists S \in \mathbf{C}(p)(S \subseteq I) \rightarrow p \in I.$$

We write $\widehat{\mathbf{C}}$ for the set of all **C**-closed sieves in P, partially ordered by inclusion.

Lemma. If $I \in \widehat{P}$, $J \in \widehat{\mathbf{C}}$, then $I \Rightarrow J \in \widehat{\mathbf{C}}$.

Proof. Suppose that $I \in \widehat{P}$, $J \in \widehat{\mathbf{C}}$, and $S \subseteq I \Rightarrow J$ with $S \in \mathbf{C}(p)$. Define $U = \{q \in I: \exists s \in S.\ q \leq s\}$. Then $U \subseteq J$. If $q \in I \cap p\downarrow$, then there is $T \in \mathbf{C}(q)$ for which $T \prec S$. Then for any $t \in T$, there is $s \in S$ for which $t \leq s$, whence $t \in U$. Accordingly $T \subseteq U \subseteq J$. Since J is a **C**-closed, it follows that $q \in J$. We conclude that $I \cap p\downarrow \subseteq J$, whence $p \in p\downarrow \subseteq I \Rightarrow J$. Therefore $I \Rightarrow J$ is **C**-closed. □

It follows from the lemma that $\widehat{\mathbf{C}}$ is a frame. For clearly an arbitrary intersection of **C**-closed sieves is **C**-closed. So $\widehat{\mathbf{C}}$ is a complete lattice. In view of the lemma the implication operation in $\widehat{P}$ restricts to one in $\widehat{\mathbf{C}}$, making $\widehat{\mathbf{C}}$ a Heyting algebra, and so a frame.

Proposition 1. Suppose that **C** is a pretopology. Then **(i)** the bottom element of $\widehat{\mathbf{C}}$ is $\mathbf{0} = \{p: \varnothing \in \mathbf{C}(p)\}$, **(ii)** the **C**-closed sieve generated by a sieve A (that is, the smallest **C**-closed sieve containing A) is $\{p: \exists S \in \mathbf{C}(p).\ S \subseteq A\}$, **(iii)** the join operation in $\widehat{\mathbf{C}}$ is given by $\bigvee_{i \in I} J_i = (\bigcup_{i \in I} J_i)^*$. If **C** is a Grothendieck topology, then **(iv)** for any sieve $S \subseteq p\downarrow$, $p \in S^* \leftrightarrow S \in \mathbf{C}(p)$.

Proof. Suppose that **C** is a pretopology. Then **0** is a **C**-closed sieve. For it is easily seen to be a sieve; and it is **C**-closed because if $S \in \mathbf{C}(p)$ and $S \subseteq \mathbf{0}$, then $\varnothing \in \mathbf{C}(s)$ for each $s \in S$, whence $\varnothing = \bigcup_{s \in S} \varnothing \in \mathbf{C}(p)$, and so $p \in \mathbf{0}$. Finally, $\mathbf{0} \subseteq I$ for any **C**-closed sieve I, for if $\varnothing \in \mathbf{C}(p)$, then from $\varnothing \subseteq I$ we infer $p \in I$. This gives **(i).** As for **(ii)**, suppose given a sieve A. Then $A \subseteq A^*$ follows from $\{p\} \in \mathbf{C}(p)$. A^* is a sieve, since if $p \in A^*$ and $q \leq p$, then there is $S \in \mathbf{C}(p)$ for which $S \subseteq A$, and $T \in \mathbf{C}(q)$ sharpening S; clearly $T \subseteq A$ also, whence $q \in A^*$. And A^* is **C**-closed, since if $S \subseteq A^*$ with $S \in \mathbf{C}(p)$, then for each $s \in S$ there is $T_s \in \mathbf{C}(s)$ with $T_s \subseteq A$; it follows that $\bigcup_{s \in S} T_s \subseteq A$ and $\bigcup_{s \in S} T_s \in \mathbf{C}(p)$, whence $p \in A^*$. Part **(iii)** is an immediate consequence of **(iii).** Finally, if **C** is a Grothendieck topology and $S \subseteq p\downarrow$ is a sieve, then $p \in S^* \leftrightarrow \exists T \in \mathbf{C}(p).\ T \subseteq S \leftrightarrow S \in \mathbf{C}(p)$, i.e. **(iv).** □

We observe parenthetically that $\widehat{\mathbf{Den}}$ *is a Boolean algebra.* To establish this it suffices to show that, for any $I \in \widehat{\mathbf{Den}}$, $\neg\neg I \subseteq I$. Now since $\varnothing \notin \mathbf{Den}(p)$, it follows from **(i)** of the proposition above that the bottom element of $\widehat{\mathbf{Den}}$ is $\varnothing$, so that, for any $I \in \widehat{\mathbf{Den}}$, $\neg I = \{p: I \cap p\downarrow = \varnothing\}$, whence $\neg\neg I = \{p: \forall q \leq p \exists r \leq q.\ r \in I\}$. But it easily checked that the defining condition for I to be a member of $\widehat{\mathbf{Den}}$ is precisely that, if $\forall q \leq p \exists r \leq q.\ r \in I$, then $p \in I$. That is, $\neg\neg I \subseteq I$.

Cover schemes on P correspond to certain self-maps on $\widehat{P}$ called (*weak*) *nuclei.* A *weak nucleus* on a frame H is a finite-meet-preserving map $j: H \rightarrow H$ such that $j(1) = 1$ and $a \leq$

$j(a)$ for any $a \in H$. If in addition $j(j(a)) \le j(a)$ (so that $j(j(a)) = j(a)$) for all $a \in H$, j is called a *nucleus* on H.

Proposition 2. Let **C** be a cover scheme on P. For each $I \in \widehat{P}$ let I^* be the least **C**-closed sieve containing I. Then the map $k_{\mathbf{C}}$: $I \mapsto I^*$ is a nucleus on $\widehat{P}$.

Proof. Clearly $I \subseteq I^*$ and $I^{**} = I^*$. It remains to be shown that, for $I, J \in \widehat{P}$, $(I \cap J)^* = I^* \cap J^*$. Since $*$ is obviously inclusion-preserving, $(I \cap J)^* \subseteq I^* \cap J^*$. For the reverse inclusion, note first that $I \in \widehat{\mathbf{C}} \leftrightarrow I^* = I$. Given $I, J \in \widehat{P}$, define $K = I \Rightarrow (I \cap J)^*$. By the Lemma above, $K \in \widehat{\mathbf{C}}$, so that $K^* = K$. Now $J^* \subseteq K$ since

$$J \cap I \subseteq (I \cap J)^* \to J \subseteq [I \Rightarrow (I \cap J)^*] = K,$$

whence $J^* \subseteq K^* \subseteq K$. Similarly, if we define $L = K \Rightarrow (I \cap J)^*$, then $I^* \subseteq L$. It follows that

$$I^* \cap J^* \subseteq K \cap L = K \cap [K \Rightarrow (I \cap J)^*] \subseteq (I \cap J)^*. \qquad \square$$

Inversely, any *weak* nucleus j on $\widehat{P}$ determines a *regular* cover scheme $\mathbf{D}_j$ on P, given by

$$S \in \mathbf{D}_j(p) \leftrightarrow p \in j(S).$$

Let us check that $\mathbf{D}_j$ is indeed a regular cover scheme. To do this it suffices to show that each $\mathbf{D}_j(p)$ is a filter of sieves and that, if $S \in \mathbf{D}_j(p)$, and $q \le p$, then $S \cap q{\downarrow} \in \mathbf{D}_j(q)$. The first of these properties follows immediately from the fact that j preserves finite intersections, and the second from the observation that, if $S \in \mathbf{D}_j(p)$, and $q \le p$, then $p \in j(S)$, so that $q \in j(S)$, and $q \in q{\downarrow} \subseteq j(q{\downarrow})$, whence $q \in j(S) \cap j(q{\downarrow}) = j(S \cap q{\downarrow})$, i.e. $S \cap q{\downarrow} \in \mathbf{D}_j(q)$.

When j is a *nucleus*, $\mathbf{D}_j$ is a *Grothendieck topology*. For under this assumption, if $S \in \mathbf{D}_j(p)$ and $T_s \in \mathbf{D}_j(s)$ for each $s \in S$, then $s \in j(T_s)$ for each $s \in S$, and it follows that

$$S \subseteq \bigcup_{s \in S} j(T_s) \subseteq j(\bigcup_{s \in S} T_s)$$

so that

$$p \in j(S) \subseteq j(j(\bigcup_{s \in S} T_s)) = j(\bigcup_{s \in S} T_s)$$

i.e., $\bigcup_{s \in S} T_s \in \mathbf{D}_j(p)$.

The correspondences $\mathbf{C} \mapsto k_{\mathbf{C}}$ and $j \mapsto \mathbf{D}_j$ between Grothendieck topologies on P and nuclei on $\widehat{P}$ are mutually inverse. For if **C** is a Grothendieck topology on P, then, by Proposition I **(iv)** we have

$$S \in \mathbf{D}_{k_{\mathbf{C}}}(p) \leftrightarrow p \in k_{\mathbf{C}}(S) = S^* \leftrightarrow S \in \mathbf{C}(p),$$

whence $\mathbf{D}_{k_{\mathbf{C}}} = \mathbf{C}$. And, for a nucleus *j* on $\widehat{P}$, we have, using Proposition 1**(ii)**,

$$\begin{aligned} k_{\mathbf{D}_j}(I) &= \text{ least } \mathbf{D}_j - \text{closed sieve } \supseteq I \\ &= \{p : \exists S \in \mathbf{D}_j(p). S \subseteq I\} \\ &= \{p : \exists S \subseteq I. p \in j(S)\} \\ &= j(I), \end{aligned}$$

whence $k_{\mathbf{D}_j} = j$.

COVER SCHEMES AND FRAMES

The relationship between cover schemes on a preordered set and (weak) nuclei on its completion can be extended to cover schemes on partially ordered sets and general frames. Accordingly let *H* be a frame: we write ı, I, ⇒ for the join, meet and implication operations, respectively, in *H.* The partially ordered set $(P, \leq)$ is said to be *dense* in *H* if *P* is a subset of *H*, the partial ordering on *P* is the restriction to *P* of that of *H*, and either of the two following equivalent conditions is satisfied: (i) for any $a \in H$, $a =$ ı $\{p: a \leq p\}$ (ii) for any $a, b \in H$, $a \leq b \leftrightarrow \forall p[p \leq a \rightarrow p \leq b]$. The canonical example of a frame in which *P* is dense is the frame $\widehat{P}$ described in section I: here each $p \in P$ is identified with the $p\downarrow \in \widehat{P}$. $\widehat{P}$ is easily seen to have the property that in it, for any $S \subseteq P$, $p \leq$ ıS iff $p \in S$.

Now fix a frame *H* in which *P* is dense and a cover scheme **C** on *P.* An element $a \in H$ is said to *cover* an element $p \in P$ if there exists a cover *S* of *p* for which ı$S \leq a$. A **C**-*element* of *H* is one which dominates every element of *P* that it covers—that is, an element $a \in H$ satisfying

$$\forall p \in P[(\exists S \in \mathbf{C}(p))2\, S \leq a \rightarrow p \leq a].$$

We write $H_{\mathbf{C}}$ for the set of all **C**-elements of *H.* It is evident that $H_{\mathbf{C}}$ is closed under the meet operation of *H.* Notice that **C**-elements and $\overline{\mathbf{C}}$-elements coincide (recalling that $\overline{\mathbf{C}}$ is the normalization of **C**.)

The canonical H-cover scheme CH on P is given by

$$S \in \mathbf{C}_H(p) \leftrightarrow \text{ı}S = p.$$

Clearly $\mathbf{C}_H$ is a pretopology, and every element of *H* is a $\mathbf{C}_H$-element.

Corresponding to the Lemma of §I, we have:

Lemma. If $a \in H, b \in H_{\mathbf{C}}$, then $a \Rightarrow b \in H_{\mathbf{C}}$.

Proof. Suppose $a \in H, b \in H_{\mathbf{C}}$, $S \in C(p)$ and ı$S \leq (a \Rightarrow b)$. Writing *U* for $\{q : q \leq a \,\&\, \exists s \in S(q \leq s)\}$, we have

$$2\,U \leq 2\,\{s \wedge q : s \in S, q \leq a\} = 2\,S \wedge 2\,\{q : q \leq a\} = 2\,S \wedge a \leq b.$$

Now if $q \leq p \wedge a$, there is $T \in \mathbf{C}(q)$ sharpening S. Then

$$t \in T \to t \leq a \;\&\; \exists s \in S(t \leq s),$$

so that $T \subseteq U$, and therefore $2\,T \leq 2\,U \leq b$. Since $b \in H_{\mathbf{C}}$, it follows that $q \leq b$. Hence $q \leq p \wedge a \to q \leq b$, so that $p \wedge a \leq b$ and $p \leq (a \Rightarrow b)$. We conclude that $(a \Rightarrow b) \in H_{\mathbf{C}}$. □

It follows from the lemma that $H_{\mathbf{C}}$ is itself a frame.

The nucleus on H *associated* with the cover scheme $\mathbf{C}$ on P is the map $j = k_{\mathbf{C}}$: $H \to H$ defined by

$$j(a) = 1\,\{x \in H_{\mathbf{C}} : a \leq x\}.$$

That j is a nucleus results from the following observations. Evidently j is order preserving, maps H onto $H_{\mathbf{C}}$, is the identity on $H_{\mathbf{C}}$, and satisfies $j(1) = 1$ and $a \leq j(a)$ for all $a \in A$. Also it is easily shown that $j(j(a)) = j(a)$. Finally, j preserves finite meets. For clearly $j(a \wedge b) \leq j(a) \wedge j(b)$ since j is order preserving. For the reverse inequality, consider first the element $u = (a \Rightarrow j(a \wedge b))$: this is, by the Lemma above, an element of $H_{\mathbf{C}}$, so that $j(u) = u$. Also $j(b) \leq u$. For from $b \wedge a \leq j(a \wedge b)$ we deduce $b \leq (a \Rightarrow j(a \wedge b)) = u$, whence $j(b) \leq j(u) = u$. Similarly, $v = ((a \Rightarrow j(a \wedge b)) \Rightarrow j(a \wedge b))$ is an element of $H_{\mathbf{C}}$ and $j(a) \leq v$. Therefore

$$j(a) \wedge j(b) \leq v \wedge u \leq j(a \wedge b),$$

as required.

Notice that the nucleus associated with a cover scheme coincides with that associated with its normalization.

Accordingly we have shown that each cover scheme on P determines a nucleus on H. Conversely, we can show that any *weak nucleus* on H determines a cover scheme on P. For, starting with a weak nucleus j on H, define the map $\mathbf{D}_j$ on P by

$$\mathbf{D}_j(p) = \{S \subseteq p\!\downarrow : p \leq j(2\,S)\}.$$

Then $\mathbf{D}_j$ is a cover scheme on P. For suppose $q \leq p$ and $S \in \mathbf{D}_j(p)$. Then $q \leq p \leq j(2\,S)$; since $q \leq j(q)$ and j preserves finite meets, it follows that

$$q \leq j(q) \wedge j(2\,S) = j(q \wedge 2\,S) = j(2\,\{s \wedge q : s \in S\}). \qquad (*)$$

Now define $T \subseteq q\!\downarrow$ by

$$T = \{t : t \leq q \;\&\; \exists s \in S(t \leq s)\}.$$

We claim that T is a ($\mathbf{D}_j$-) cover of q sharpening S. That T sharpens S is evident from its definition. To see that it is a cover of q we observe that, if $s \in S$, then

$$s \wedge q = 2\,\{t : t \le s \wedge q\} = 2\,\{t : t \le s \,\&\, t \le q\} \le 2\, T.$$

Therefore $2\,\{s \wedge q : s \in S\} \le 2\, T$, so that, by (*),

$$q \le j(2\,\{s \wedge q : s \in S\}) \le j(2\, T),$$

that is, T covers q.

When j is a *nucleus,* the associated cover scheme $\mathbf{D}_j$ is actually a *pretopology*. For in any case $\{p\} \in \mathbf{D}_j(p)$. Moreover, if j is a nucleus, $S \in \mathbf{D}_j(p)$ and $T_s \in \mathbf{D}_j(s)$ for each $s \in S$, then

$$p \le j(2\, S) \le j(\underset{s \in S}{2}\, j(2\, T_s)) \le j(j(\underset{s \in S}{2}\, 2\, T_s) = j(2 \bigcup_{s \in S} T_s).$$

Therefore $\bigcup_{s \in S} T_s \in \mathbf{D}_j(p)$, and $\mathbf{D}_j$ is a pretopology.

Starting with a weak nucleus j, we obtain the corresponding cover scheme $\mathbf{D}_j$. The latter in turn determines a nucleus j^* given by

$$j^*(a) = 1\,\{x \in H_{\mathbf{D}_j} : a \le x\}.$$

Now by definition, we have

$$\begin{aligned} x \in H_{\mathbf{D}_j} &\leftrightarrow \forall p[(\exists S \in \mathbf{D}_j(p)) 2\, S \le x \to p \le x] \\ &\leftrightarrow \forall p[(\exists S \subseteq p\!\downarrow)(p \le j(2\, S) \,\&\, 2\, S \le x) \to p \le x] && \text{(a)} \\ &\leftrightarrow \forall p[p \le j(x) \to p \le x] && \text{(b)} \\ &\leftrightarrow j(x) = x. \end{aligned}$$

(To see the equivalence between (a) and (b), we need to establish the equivalence between $(\exists S \subseteq p\!\downarrow)(p \le j(2\, S) \,\&\, 2\, S \le x)$ and $p \le j(x)$. Clearly the first of these implies the second. As for the converse, if $p \le j(x)$, then since $p \le j(p)$, we have

$$p \le j(x) \wedge j(p) = j(x \wedge p) = j(2\, S),$$

where $S = \{q : q \le x \wedge p\}$. Then $S \subseteq p\!\downarrow$, $p \le j(\imath S)$ and $\imath S \le x$, and the first statement follows.) Accordingly

$$j^*(a) = 1\,\{x \in H : a \le x \,\&\, j(x) = x\}. \qquad (*)$$

j^* is called the nucleus[3] *generated* by the weak nucleus j; it is easily deduced from (*) that when j is a nucleus, j^* and j coincide.

The generation of nuclei by weak nuclei can itself be seen as an instance of a nuclear operation. For consider the set $W(H)$ of all weak nuclei on H. When $W(H)$ is partially ordered pointwise in the obvious way, it becomes a frame with implications, joins, and meets given by the following specifications: $(j \Rightarrow k)(a) = \underset{b \geq a}{1} (j(b) \Rightarrow k(b))$ and for $S \subseteq W(H)$,

$(2\, S)(a) = \underset{s \in S}{2}\; s(a)$, $(1\, S)(a) = \underset{s \in S}{1}\; s(a)$. The subset $N(H)$ of $W(H)$ consisting of all nuclei can be shown to be a sublocale (see [3]) of $N(H)$, that is, it is closed under arbitrary meets in $W(H)$ and is such that $(j \Rightarrow k) \in N(H)$ whenever $j \in W(H)$, $k \in N(H)$. That being the case, the map φ: $W(H) \to N(H)$ defined by

$$\varphi(j) = 1\, \{k \in N(H) : j \leq k\}$$

is a nucleus on $W(H)$, and it is easily shown that $\varphi(j) = j^*$. So the generation of nuclei by weak nuclei is precisely the action of the nucleus φ.

Now start with a cover scheme $\mathbf{C}$ on P, obtain the associated nucleus $k_{\mathbf{C}}$ on H, and consider its associated cover scheme $\mathbf{D}_{k_{\mathbf{C}}} = \mathbf{C}^*$ on P. By definition we have, for $S \subseteq p\!\downarrow$,

$$\begin{aligned} S \in \mathbf{C}^*(p) &\leftrightarrow p \leq j_{\mathbf{C}}(2\, S) \\ &\leftrightarrow p \leq 1\, \{x \in H_{\mathbf{C}} : 2\, S \leq x\} \\ &\leftrightarrow \forall x \in H_{\mathbf{C}}(2\, S \leq x \to p \leq x) \end{aligned}$$

Recalling the definition of $H_{\mathbf{C}}$, we see immediately that this last assertion is implied by $S \in \mathbf{C}(p)$, so that always $\mathbf{C}(p) \subseteq \mathbf{C}^*(p)$. The reverse inclusion will hold, and so $\mathbf{C}$ will coincide with $\mathbf{C}^*$, precisely when the cover scheme $\mathbf{C}$ is *saturated*, that is, coincides with its *saturate*, which we next proceed to define.

The (H-) *saturate* $\widetilde{\mathbf{C}}$ of a cover scheme C on P is defined by

$$\widetilde{\mathbf{C}}(p) = \{S \subseteq p\!\downarrow : \forall x \in H_{\mathbf{C}}(2\, S \leq x \to p \leq x)\}.$$

Then $\widetilde{\mathbf{C}}$ is a cover scheme. For if $S \in \widetilde{\mathbf{C}}(p)$ and $q \leq p$, consider the subset T of $p\!\downarrow$ defined by

$$T = \{t \leq q : \exists s \in S(t \leq s)\}.$$

It is easily shown that $2\, T = (2\, S) \wedge q$. Now if $x \in H_{\mathbf{C}}$ and $1T \leq x$, then $1S \wedge q \leq x$, whence $1S \leq (q \Rightarrow x)$. But since x is an element of $H_{\mathbf{C}}$, so, by the lemma, is $q \Rightarrow x$, and since $S \in \widetilde{\mathbf{C}}(p)$, it follows that $p \leq (q \Rightarrow x)$. Thus $q = p \wedge q \leq x$. Accordingly $T \in \widetilde{\mathbf{C}}(q)$, and T obviously sharpens S. This shows that $\widetilde{\mathbf{C}}$ is indeed a cover scheme.

[3] It can be verified directly that j^* is a nucleus.

It is readily shown that any cover scheme associated with a nucleus (as opposed to a weak nucleus) is saturated. Observe that, when H is $\widehat{P}$, every coverage on P is saturated, since in that case $H_{\mathbf{C}}$ is $\widehat{\mathbf{C}}$ and so we have, using Proposition I.1 (**iv**),

$$S \in \widetilde{\mathbf{C}}(p) \leftrightarrow \forall I \in \widehat{\mathbf{C}}[S \subseteq I \to p \in I] \leftrightarrow p \in S^* \leftrightarrow S \in \mathbf{C}(p).$$

To sum up, each weak nucleus on H gives rise to a cover scheme on P and the cover scheme associated with a nucleus is saturated. Conversely, each cover scheme gives rise to a nucleus. This establishes mutually inverse correspondences between nuclei and saturated cover schemes.

Given $a \in H$, we define the nuclei $j_a\ j^a$ on H by

$$j_a(x) = a \vee x \qquad j^a(x) = a \Rightarrow x.$$

The associated cover schemes (easily seen to be pretopologies) on P are given by:

$$S \in \mathbf{C}_a(p) \leftrightarrow p \leq a \vee 2\, S$$
$$S \in \mathbf{C}^a(p) \leftrightarrow p \wedge a \leq 2\, S.$$

Notice that

$$p \leq a \leftrightarrow \varnothing \in \mathbf{C}_a(p)$$
$$p \leq \neg a \leftrightarrow \varnothing \in \mathbf{C}^a(p).$$

The *double negation operation* $\neg\neg$ is a nucleus on H, whose associated cover scheme is precisely the dense cover scheme **Den** (which accordingly is also known as the *double negation cover scheme*). An argument similar to the one above showing that $\widehat{\mathbf{Den}}$ is a Boolean algebra establishes that $H_{\mathbf{Den}}$ is a Boolean algebra: it is in fact the complete Boolean algebra of $\neg\neg$-closed elements of H.

Proposition. Let j be a weak nucleus on H. Then the following are equivalent: (a) $j(0) = 0$ (b) $j \leq \neg\neg$ (in the pointwise ordering of $W(H)$) (c) $\varnothing \notin \mathbf{D}_j(p)$ for all p.

Proof. If $j \leq \neg\neg$ then $j0 \leq \neg\neg 0 = 0$. Conversely if $j0 = 0$ then, for any $a \in H$, $j(a) \wedge \neg a \leq j(a) \wedge j(\neg a) = j(a \wedge \neg a) = j0 = 0$. So $j(a) \leq \neg\neg a$. Finally, we have

$$\begin{aligned} j0 = 0 \leftrightarrow 0 \in H_{\mathbf{D}_j} &\leftrightarrow \forall p[(\exists S \in \mathbf{D}_j(p))2\, S = 0 \to p = 0] \\ &\leftrightarrow \forall p[\neg(\exists S \in \mathbf{D}_j(p)2\, S = 0] \\ &\leftrightarrow \forall p[\varnothing \notin \mathbf{D}_j(p)]. \qquad \square \end{aligned}$$

Cover Schemes and Kripke Models

We revert to the assumption that $(P, \leq)$ is a preordered set. Recall that a *presheaf* on P is an assignment, to each $p \in P$, of a set $\mathbf{F}(p)$ and to each pair (p, q) with $q \leq p$ of a map $\mathbf{F}_{pq}$: $\mathbf{F}(p) \to \mathbf{F}(q)$ in such a way that $\mathbf{F}_{pp}$ is the identity on $\mathbf{F}(p)$ and, for $r \leq q \leq p$, $\mathbf{F}_{pr} = \mathbf{F}_{qr} \quad \mathbf{F}_{pq}$. The set $V(\mathbf{F}) = \bigcup_{p \in P} \mathrm{F}(p)$ is called the *universe* of $\mathbf{F}$. A *Kripke model* based on P is a presheaf K for which $\mathrm{K}(p) \subseteq \mathrm{K}(q)$ whenever $q \leq p$ and each $\mathbf{K}_{pq}$ is the corresponding insertion map. Put more simply, a Kripke model based on P is a map $\mathbf{K}$ from P to a family of sets satisfying $\mathrm{K}(p) \subseteq \mathrm{K}(q)$ whenever $q \leq p$. A Kripke model K based on P may be regarded as a set "evolving" or "growing" over P: each $\mathrm{K}(p)$ may be thought of as the "state" of the evolving set K at "stage" p.

Now suppose that we are given a cover scheme $\mathbf{C}$ on P. A Kripke model $\mathbf{K}$ based on P satisfying

$$\mathbf{K}(p) = \bigcap_{s \in S} \mathbf{K}(s)$$

for any $p \in P$, $S \in \mathbf{C}(p)$ is said to be *compatible with* $\mathbf{C}$. (When P is directed downward, that is, whenever each pair of elements of P has a lower bound, and $\mathbf{C}$ is a pretopology on P, a Kripke model compatible with $\mathbf{C}$ is nothing other than a $\mathbf{C}$-*sheaf*.)

Each Kripke model K based on P induces a Kripke model $\widetilde{\mathrm{K}}$ based on the free lower semilattice $\widetilde{P}$ generated by P by setting

$$\widetilde{\mathrm{K}}(\varnothing) = \varnothing \qquad \widetilde{\mathrm{K}}(\{p_1,\ldots p_n\}) = \mathrm{K}(p_1) \cup \ldots \cup \mathrm{K}(p_n).$$

If, further, K is compatible with the cover scheme $\mathbf{C}$ on P, then $\widetilde{\mathrm{K}}$ is compatible with the cover scheme on $\widetilde{P}$ induced by P (and hence also with the associated coverage on $\widetilde{P}$.) For suppose that K is in fact compatible with the cover scheme $\mathbf{C}$ on P. Given $F \in \widetilde{P}$, a nonempty subset $\{p_1, \ldots, p_n\}$ of F, and $S_1 \in \mathbf{C}(p_1), \ldots, S_n \in \mathbf{C}(p_n)$, we have

$$\begin{aligned}\bigcap_{X \in S_1 \bullet \ldots \bullet S_n} \widetilde{K}(X) &= \bigcap_{s_1 \in S_1, \ldots, s_n \in S_n} \widetilde{K}(\{s_1, \ldots, s_n\} \cup F) \\ &= \bigcap_{s_1 \in S_1, \ldots, s_n \in S_n} (K(s_1) \cup \ldots \cup K(s_n) \cup \bigcup_{p \in F} K(p)) \\ &= \bigcap_{s_1 \in S_1} K(s_1) \cup \ldots \cup \bigcap_{s_n \in S_n} K(s_n) \cup \bigcup_{p \in F} K(p)) \\ &= K(p_1) \cup \ldots \cup K(p_n) \cup \bigcup_{p \in F} K(p) \\ &= \widetilde{K}(F).\end{aligned}$$

Now suppose that the cover scheme **C** is in fact a *pretopology*. Then any Kripke model K based on P induces a Kripke model K_C also based on P but in addition compatible with **C** given by

$$K_{\mathbf{C}}(p) = \bigcup_{S \in \mathbf{C}(p)} \bigcap_{s \in S} K(s),$$

that is,

$$a \in K_{\mathbf{C}}(p) \leftrightarrow \exists S \in \mathbf{C}(p) \forall s \in S.\ a \in K(s).$$

We note that $K(p) \subseteq K_C(p)$ for every p. It is easily checked that this defines a Kripke model over P; let us confirm its compatibility with **C.** It suffices to show that, given $S \in \mathbf{C}(p)$, we have $\bigcap_{s \in S} K_{\mathbf{C}}(s) \subseteq K_{\mathbf{C}}(p)$. Indeed, if $a \in \bigcap_{s \in S} K_{\mathbf{C}}(s)$,. then for each $s \in S$ there is $T_s \in \mathbf{C}(s)$ with $a \in \bigcap_{t \in T_s} K(t)$. Writing $T = \bigcup_{s \in S} T_s$, we then have $T \in \mathbf{C}(p)$ and $a \in \bigcap_{t \in T} K(t)$. It follows that $a \in K_{\mathbf{C}}(p)$, as required.

Let us now examine some special cases. Let U be a subset of the universe V of K, and let U^* be the sieve $\{p\colon U \subseteq K(p)\}$. Now consider the Kripke model K^U compatible with $\mathbf{C}^{U^*}$ induced by K. For arbitrary $p \in P$, we have $S = p{\downarrow} \cap U^* \in \mathbf{C}^{U^*}(p)$. and $U \subseteq K(s) \subseteq K^U(s)$ for every $s \in S$. Hence $U \subseteq \bigcap_{s \in S} K^U(s) = K^U(p)$. Thus, under these conditions, U is a subset of *every* $K^U(p)$. In other words, the passage from K to K^U *forces* U to be included in the state of K^U at each stage. (Note that if $U^* = \varnothing$ then K^U assumes the constant value V.)

Again let U be a subset of V; this time define U^+ to be the sieve $\{p\colon U \cap K(p) \neq \varnothing\}$. Now consider the Kripke model K_U compatible with $\mathbf{C}_{U+}$ induced by K. Then for any p we have

$$U \cap K(p) \neq \varnothing \rightarrow p \in U^+ \rightarrow \varnothing \in \mathbf{C}_{U+}(p) \rightarrow \mathbf{K}_U(p) = \bigcap_{s \in \varnothing} K_U(s) = V.$$

That is, the passage from **K** to $\mathbf{K}_U$ *forces* each state of K_U, apart from those already maximal, to be disjoint from *U*. (Notice that if $U^+ = P$, then K_U assumes the constant value *V*.)

We next turn to *logic* in Kripke models. Each Kripke model **K** based on *P*, with universe *V*, determines a map $\widehat{\mathsf{K}} : V \to \widehat{P}$ given by

$$\widehat{\mathsf{K}}(v) = \{p : v \in \mathsf{K}(p)\}.$$

This extends naturally to a homomorphism—also denoted by $\widehat{\mathsf{K}}$ —of the free Heyting algebra *F*(*V*) generated by *V* into $\widehat{P}$. Think of the members of *F*(*V*) as the formulas of intuitionistic propositional logic generated by the members of *V* regarded as propositional atoms. Introduce the familiar *forcing* relation $\Vdash_{\mathbf{K}}$ between *P* and *F*(*V*) by defining

$$p \Vdash_{\mathbf{K}} \varphi \;\leftrightarrow\; p \in \widehat{\mathsf{K}}(\varphi). \tag{*}$$

Then the fact that $\widehat{\mathsf{K}} : F(V) \to \widehat{P}$ is a homomorphism of Heyting algebras translates into the usual rules for "Kripke semantics", namely

- $p \Vdash_{\mathbf{K}} \varphi \wedge \psi \leftrightarrow p \Vdash_{\mathbf{K}} \varphi \;\;\&\;\; p \Vdash_{\mathbf{K}} \psi$
- $p \Vdash_{\mathbf{K}} \varphi \vee \psi \leftrightarrow p \Vdash_{\mathbf{K}} \varphi \;\text{ or }\; p \Vdash_{\mathbf{K}} \psi$
- $p \Vdash_{\mathbf{K}} \varphi \Rightarrow \psi \;\leftrightarrow\; \forall q {\leq} p[\, q \Vdash_{\mathbf{K}} \varphi \;\to\; q \Vdash_{\mathbf{K}} \psi]$
- $p \Vdash_{\mathbf{K}} \neg\varphi \;\leftrightarrow\; \forall q {\leq} p \;\; q \nVdash_{\mathbf{K}} \varphi$

Equally, the map $\widehat{\mathsf{K}} : V \to \widehat{P}$ extends to a frame homomorphism (i.e., a map preserving top elements, ∧, and ı)—again denoted by $\widehat{\mathsf{K}}$ —of the *free frame* Φ(*V*) generated by *V*. Think of the members of Φ(*V*) as the formulas of *infinitary* intuitionistic propositional logic generated by the members of *V* regarded as propositional atoms. Such a formula φ is said to be *geometric* if it is generated from propositional atoms by applying just ∧ and ı. Introducing the forcing relation $\Vdash_{\mathbf{K}}$ between *P* and Φ(*V*) as in (*) above, the fact that $\widehat{\mathsf{K}} : \Phi(V) \to \widehat{P}$ is a frame homomorphism translates into the semantical rules for *geometric* formulas:

- $p \Vdash_{\mathbf{K}} \varphi \;\wedge\; \psi \;\;\leftrightarrow\;\; p \Vdash_{\mathbf{K}} \varphi \;\;\&\; p \Vdash_{\mathbf{K}} \psi$
- $p \Vdash_{\mathbf{K}} \underset{i\in I}{2} \varphi_i \;\;\leftrightarrow\;\; \exists i \in I \;\; p \Vdash_{\mathbf{K}} \varphi_i$.

Now suppose that **C** is a pretopology on *P*. It is then easily seen that **K** is compatible with **C** iff each $\widehat{\mathsf{K}}(v)$ is a **C**-closed sieve. So if **K** is compatible with **C**, the resulting map $\widehat{\mathsf{K}} : V \to \widehat{\mathbf{C}}$ can be extended to a homomorphism, which we shall denote by $\widehat{\mathsf{K}}_{\mathbf{C}}$, of *F*(*V*) into $\widehat{\mathbf{C}}$. Introducing the forcing relation $\Vdash_{\mathsf{K},\mathbf{C}}$ between *P* and *F*(*V*) by

$$p \Vdash_{K,\mathbf{C}} \varphi \leftrightarrow p \in \widehat{K_{\mathbf{C}}}(\varphi), \tag{**}$$

we find that the fact that $\widehat{K}_{\mathbf{C}} : F(V) \to \widehat{\mathbf{C}}$ is a homomorphism translates into the rules of "Beth-Kripke-Joyal" semantics for $\Vdash_{K,\mathbf{C}}$ (see, e.g., [4]), viz.,

- $p \Vdash_{K,\mathbf{C}} \varphi \wedge \psi \leftrightarrow p \Vdash_{K,\mathbf{C}} \varphi \;\&\; p \Vdash_{K,\mathbf{C}} \psi$
- $p \Vdash_{K,\mathbf{C}} \varphi \vee \psi \leftrightarrow \exists S \in \mathbf{C}(p)\, \forall s \in S\, [s \Vdash_{K,\mathbf{C}} \varphi$ or $s \Vdash_{K,\mathbf{C}} \psi]$
- $p \Vdash_{K,\mathbf{C}} \varphi \Rightarrow \psi \leftrightarrow \forall q \le p [\, q \Vdash_{K,\mathbf{C}} \varphi \rightarrow q \Vdash_{K,\mathbf{C}} \psi]$
- $p \Vdash_{K,\mathbf{C}} \neg\varphi \leftrightarrow \forall q \le p\, [q \Vdash_{K,\mathbf{C}} \varphi \rightarrow \varnothing \in \mathbf{C}(p)]$.

We verify the second and fourth of these. We have, using Proposition I 1. (iii),

$$p \Vdash_{K,\mathbf{C}} \varphi \vee \psi \leftrightarrow p \in \widehat{K_{\mathbf{C}}}(\varphi \vee \psi) = \widehat{K_{\mathbf{C}}}(\varphi) \vee \widehat{K_{\mathbf{C}}}(\psi) \text{ (in } \widehat{\mathbf{C}})$$

$$\leftrightarrow \exists S \in C(p).\; S \subseteq \widehat{K_{\mathbf{C}}}(\varphi) \cup \widehat{K_{\mathbf{C}}}(\psi)$$

$$\leftrightarrow \exists S \in C(p) \forall s \in S.\; s \in \widehat{K_{\mathbf{C}}}(\varphi) \vee s \in \widehat{K_{\mathbf{C}}}(\psi)$$

$$\leftrightarrow \exists S \in \mathbf{C}(p)\, \forall s \in S\, [s \Vdash_{K,\mathbf{C}} \varphi \text{ or } s \Vdash_{K,\mathbf{C}} \psi].$$

Also, using Proposition I 1. (ii) we have

$$p \Vdash_{K,\mathbf{C}} \neg\varphi \leftrightarrow p \in \widehat{K_{\mathbf{C}}}(\neg\varphi) = \neg\widehat{K_{\mathbf{C}}}(\varphi) = (\widehat{K_{\mathbf{C}}}(\varphi) \Rightarrow \mathbf{0})$$

$$\leftrightarrow \forall q \le p [q \in \widehat{K}_{\mathbf{C}}(\varphi) \rightarrow \varnothing \in \mathbf{C}(q)]$$

$$\leftrightarrow \forall q \le p [q \Vdash_{K,\mathbf{C}} \varphi \rightarrow \varnothing \in \mathbf{C}(q)].$$

Since $\widehat{\mathbf{Den}}$ is a Boolean algebra it follows that, when K is compatible with **Den**, $p \Vdash_{K,\mathbf{Den}} \varphi \vee \neg\varphi$ for every p, i.e., classical logic prevails in the Kripke model associated with $\widehat{K}_{\mathbf{Den}}$.

When K is compatible with **C**, the map $\widehat{K} : V \to \widehat{\mathbf{C}}$ can be extended to a frame homomorphism, which we shall again denote by $\widehat{K}_{\mathbf{C}}$, of $\Phi(V)$ into $\widehat{\mathbf{C}}$. Introduce the forcing relation $\Vdash_{K,\mathbf{C}}$, now between P and $\Phi(V)$, by the same equivalence (**) as above. When **C** is a Grothendieck topology, a straightforward inductive argument shows that, for any geometric formula φ,

$$p \Vdash_{K,\mathbf{C}} \varphi \leftrightarrow \exists S \in \mathbf{C}(p)\, \forall s \in S\, .\, s \Vdash_K \varphi. \qquad (\dagger)$$

This may be applied to "force" any given set Σ of geometric formulas to become true in a Kripke model. For, starting with a Kripke model K, let A be the sieve $\{p\colon \forall\sigma\in\Sigma.\, p \Vdash_K \sigma\}$. Let $\mathbf{G}$ be the Grothendieck topology generated by the coverage $\mathbf{C}^A$: it is easily verified that a sieve $S \subseteq p{\downarrow}$ satisfies the same condition for membership in $\mathbf{G}(p)$ as in $\mathbf{C}^A(p)$, viz., $p{\downarrow} \cap A \subseteq S$. Now by (†) we have, for each $\sigma \in \Sigma$,

$$p \Vdash_{K,\mathbf{G}} \sigma \leftrightarrow \exists S \in \mathbf{G}(p)\, \forall s \in S\, .\, s \Vdash_K \sigma. \qquad (\ddagger)$$

If we take S to be $p{\downarrow} \cap A$, then evidently $S \in \mathbf{G}(p)$ and $\forall s\in S\, .\, s \Vdash_K \sigma$. It now follows from (‡) that $p \Vdash_{K,\mathbf{G}} \sigma$ for every $\sigma \in \Sigma$ and every $p \in P$. In this sense $\mathbf{G}$ "forces" all the members of Σ to be true in the Kripke model associated with $\widehat{K}_{\mathbf{G}}$.

COVER SCHEMES AND FRAME-VALUED SET THEORY

We now set about relating what has been done so far to frame-valued set theory. Associated with each frame H is an H-valued model $V^{(H)}$ of (intuitionistic) set theory (see, e.g. [1] or [2]): we recall some of its principal features.

- Each of the members of $V^{(H)}$—the *H-valued sets*—is a map from a subset of $V^{(H)}$ to H.
- Corresponding to each sentence σ of the language of set theory (with names for all elements of $V^{(H)}$) is an element $[\![\sigma]\!] = [\![\sigma]\!]^H \in H$ called its *truth value in* $V^{(H)}$. These "truth values" satisfy the following conditions. For $a, b \in V^{(H)}$,

$$[\![b \in a]\!] = \underset{c\in dom(a)}{2} [\![b = c]\!] \wedge a(c)$$

$$[\![b = a]\!] = \underset{c\in dom(a)\cup dom(b)}{2} ([\![c \in b]\!] \Leftrightarrow [\![c \in a]\!])$$

$$[\![\sigma \wedge \tau]\!] = [\![\sigma]\!] \wedge [\![\tau]\!], \text{ etc.}$$

$$[\![\exists x\varphi(x)]\!] = \underset{a\in V^{(H)}}{2} [\![\varphi(a)]\!]$$

$$[\![\forall x\varphi(x)]\!] = \underset{a\in V^{(H)}}{1} [\![\varphi(a)]\!]$$

A sentence σ is *valid,* or *holds,* in $V^{(H)}$, written $V^{(H)} \vDash \sigma$, if $[\![\sigma]\!] = 1$, the top element of H. The truth value $[\![\sigma]\!]$ "measures" the degree or extent to which σ holds: the larger $[\![\sigma]\!]$ is, the "truer" σ is. In particular, when $[\![\sigma]\!] = 1$, σ is 'universally" or "absolutely" true, and when $[\![\sigma]\!] = 0$, σ is "universally" or "absolutely" false.

- The axioms of intuitionistic Zermelo-Fraenkel set theory are valid in $V^{(H)}$. Accordingly the category $\mathtt{Set}^{(H)}$ of sets constructed within $V^{(H)}$ is a topos: in fact $\mathtt{Set}^{(H)}$ can be shown to be equivalent to the topos of canonical sheaves on H.
- There is a canonical embedding $x \mapsto \hat{x}$ of the universe V of sets into $V^{(H)}$ satisfying

$$[\![u \in \hat{x}]\!] = \underset{y \in x}{2}\ [\![u = \hat{y}]\!] \quad \text{for } x \in V, u \in V^{(H)}$$

$$x \in y \leftrightarrow V^{(H)} \vDash \hat{x} \in \hat{y}, \quad x = y \leftrightarrow V^{(H)} \quad \hat{x} = \hat{y} \quad \text{for } x, y \in V$$

$$\varphi(x_1, \ldots, x_n) \leftrightarrow V^{(H)} \vDash \varphi(x_1, \ldots, x_n) \text{ for } x_1, \ldots, x_n \in V \text{ and restricted } \varphi$$

(Here a formula φ is *restricted* if all its quantifiers are restricted, i.e. can be put in the form $\forall x \in y$ or $\exists x \in y$.)

It follows from the last of these assertions that the canonical representative $\widehat{H}$ of H is a Heyting algebra in $V^{(H)}$. The *canonical prime filter* in $\widehat{H}$ is the H-set Φ_H defined by

$$dom(\Phi_H) = \{\hat{a} : a \in H\}, \quad \Phi_H(\hat{a}) = a \quad \text{for } a \in H.$$

Clearly $V^{(H)} \vDash \Phi_H \subseteq \widehat{H}$, and it is easily verified that

$$V^{(H)} \vDash \Phi_H \text{ is a (proper) prime filter}^4 \text{ in } \widehat{H}.$$

It can also be shown that Φ_H is *V-generic* in the sense that, for any subset $A \subseteq H$,

$$V^{(H)} \vDash \widehat{2A} \in \Phi_H \leftrightarrow \Phi_H \cap \hat{A} \neq \varnothing.$$

Moreover, for any $a \in H$ we have $[\![\hat{a} \in \Phi_H]\!] = a$, and in particular, for any sentence σ, $[\![\sigma]\!] = [\![\widehat{[\![\sigma]\!]} \in \Phi_H]\!]$. Thus $V^{(H)} \vDash \sigma \leftrightarrow V^{(H)} \vDash \widehat{[\![\sigma]\!]} \in \Phi_H$—in this sense Φ_H is the filter of "true" sentences in $V^{(H)}$.

This suggests that we define a *truth set* in $V^{(H)}$ to be an H-set F for which

$$V^{(H)} \vDash F \text{ is a filter in } \widehat{H} \text{ such that } F \supseteq \Phi_H.$$

There is a natural bijective correspondence between truth sets in $V^{(H)}$ and weak nuclei on H. With each weak nucleus j on H we associate the H-set T_j defined by $\mathrm{dom}(T_j) = \mathrm{dom}(\Phi_H)$ and $T_j(\hat{a}) = j(a)$ for $a \in H$. It is easily verified that T_j is a truth set—the requirement that T_j be a filter corresponds exactly to the condition that j preserve finite meets and that it contain

4 We recall that a filter F in a lattice is *prime* if $x \vee y \in F$ implies $x \in F$ or $y \in F$.

Φ_H to the condition that j satisfy $a \leq j(a)$. Inversely, given a truth set F in $V^{(H)}$, we define the map $j_F : H \to H$ by $j_F(a) = [\![\hat{a} \in T]\!]$. Again, it is readily verified that j_F is a weak nucleus on H. These correspondences are evidently mutually inverse and in fact establish an isomorphism between the frame $W(H)$ of weak nuclei on H and the internal frame of filters in $\widehat{H}$ containing Φ_H. Under this isomorphism *nuclei* correspond precisely to *reflexive truth sets,* that is, truth sets satisfying the additional condition (evidently met by Φ_H)

$$V^{(H)} \models \widehat{[\![\hat{a} \in F]\!]} \in F \to \hat{a} \in F.$$

It is of interest to examine the familiar case in which H is a complete *Boolean algebra B.* In this case the canonical prime filter Φ_B is an *ultrafilter* in $\widehat{B}$, so that, in $V^{(B)}$, the only filters in $\widehat{B}$ containing Φ_B—the only truth sets—are Φ_B itself and $\widehat{B}$. It follows that, for truth sets F and G in $V^{(B)}$

$$V^{(B)} \models F = G \leftrightarrow [\hat{0} \in F \leftrightarrow \hat{0} \in G].$$

In other words, the truth value $[\![\hat{0} \in F]\!]$, which can be an arbitrary member of B, determines the identity of F. This means that truth sets in $V^{(B)}$, and so equally weak nuclei on B, are in bijective correspondence with the members of B. In fact it is readily shown directly that any weak nucleus on a Boolean algebra B is of the form j_a for some $a \in B$. For given a weak nucleus j on B, observe: $\neg x \leq j(\neg x)$, whence $\neg j(\neg x) \leq \neg\neg x = x$. Also $j(x) \wedge j(\neg x) = j(x \wedge \neg x) = j(0)$, whence $j(x) \leq j(\neg x) \Rightarrow j(0) = \neg j(\neg x) \vee j(0) \leq x \vee j(0)$. But clearly $x \vee j(0) \leq j(x)$, so that $j(x) = x \vee j(0)$.

Consider now the special case in which H is the completion $\widehat{P}$ of a preordered set P. We have already established a bijective correspondence between Grothendieck topologies on P and nuclei on $\widehat{P}$. This leads in turn to a bijective correspondence between Grothendieck topologies on P and reflexive truth sets in $V^{(\widehat{P})}$. Explicitly, this correspondence assigns to each Grothendieck topology $\mathbf{C}$ on P the reflexive truth set $T_{\mathbf{C}}$ in $V^{(\widehat{P})}$ given by $T_{\mathbf{C}}(S) = S^*$ for $S \in \widehat{P}$, and to each reflexive truth set F in $V^{(\widehat{P})}$ the Grothendieck topology $\mathbf{C}_F$ on P defined by $S \in \mathbf{C}_T(p) \leftrightarrow p \in [\![\hat{S} \in T]\!]$.

The topos $\mathbf{Set}^{(\widehat{P})}$ of sets in $V^{(\widehat{P})}$ is, as we have observed, equivalent to the topos of canonical sheaves on $\widehat{P}$, which is itself, as is well known, equivalent to the topos $\mathbf{Set}^{P^{op}}$ of presheaves on P. Moreover, Grothendieck topologies on P are known (see [4]) to correspond bijectively to internal Lawvere-Tierney topologies—that is, internal nuclei—on the truth-value object Ω in $\mathbf{Set}^{P^{op}}$. How this fact related to the representation of Grothendieck topologies as reflexive truth sets in $V^{(\widehat{P})}$? It turns out that in a general $V^{(H)}$ there is a natural bijection between truth sets/reflexive truth sets and weak nuclei/nuclei on $\Omega = \{u \colon u \subseteq \hat{1}\}$.

The representation of Grothendieck topologies as truth sets in $V^{(H)}$, while equivalent to that through Lawvere-Tierney topologies, seems especially perspicuous.

The *forcing* relation $\Vdash_P$ in $V^{(\widehat{P})}$ between sentences and elements of P is defined by

$$p \Vdash_P \sigma \leftrightarrow p \in [\![\sigma]\!]^{\widehat{P}}.$$

Note that we then have

$$[\![\sigma]\!]^{\widehat{P}} = \{p : p \Vdash_P \sigma\}.$$

$\Vdash_P$ satisfies the usual rules governing Kripke semantics for predicate sentences, viz.,

- $p \Vdash_P \varphi \wedge \psi \leftrightarrow p \Vdash_P \varphi$ & $p \Vdash_P \psi$
- $p \Vdash_P \varphi \vee \psi \leftrightarrow p \Vdash_P \varphi$ or $p \Vdash_P \psi$
- $p \Vdash_P \varphi \rightarrow \psi \leftrightarrow \forall q \leq p[\, q \Vdash_P \varphi \rightarrow q \Vdash_P \psi]$
- $p \Vdash_P \neg\varphi \leftrightarrow \forall q \leq p \;\, q \nVdash_K \varphi$
- $p \Vdash_P \forall x\, \varphi \leftrightarrow p \Vdash_P \varphi(a)$ for every $a \in V^{(\widehat{P})}$
- $p \Vdash_P \exists x\, \varphi \leftrightarrow p \Vdash_P \varphi(a)$ for some $a \in V^{(\widehat{P})}$.

We note also that $\Vdash_P$ is *persistent* in the sense that, if $p \Vdash_P \varphi$ and $q \leq p$, then $q \Vdash_P \varphi$.

If $\mathbf{C}$ be a pretopology on P, the forcing relation $\Vdash_{\mathbf{C}}$ in the model $V^{(\widehat{\mathbf{C}})}$ is similarly defined by

$$p \Vdash_{\mathbf{C}} \sigma \leftrightarrow p \in [\![\sigma]\!]_{\widehat{\mathbf{C}}}.$$

As for Kripke models, this relation can be shown to satisfy the rules of Beth-Kripke-Joyal semantics, viz.,

- $p \Vdash_{\mathbf{C}} \varphi \wedge \psi \leftrightarrow p \Vdash_{\mathbf{C}} \varphi$ & $p \Vdash_{\mathbf{C}} \psi$
- $p \Vdash_{\mathbf{C}} \varphi \vee \psi \leftrightarrow \exists S \in \mathrm{C}(p)\, \forall s \in S\; [s \Vdash_{\mathbf{C}} \varphi$ or $s \Vdash_{\mathbf{C}} \psi]$
- $p \Vdash_{\mathbf{C}} \varphi \Rightarrow \psi \leftrightarrow \forall q \leq p[\, q \Vdash_{\mathbf{C}} \varphi \rightarrow q \Vdash_{\mathbf{C}} \psi]$
- $p \Vdash_{\mathbf{C}} \neg\varphi \leftrightarrow \forall q \leq p\; [q \Vdash_{\mathbf{C}} \varphi \rightarrow \varnothing \in \mathrm{C}(p)]$
- $p \Vdash_{\mathbf{C}} \forall x\varphi \leftrightarrow p \Vdash_{\mathrm{C}} \varphi(a)$ for every $a \in V^{(\widehat{\mathbf{C}})}$
- $p \Vdash_{\mathbf{C}} \exists x\varphi \leftrightarrow \exists S \in \mathrm{C}(p)\, \forall s \in S\;\, s \Vdash_{\mathrm{C}} \varphi(a)$ for some $a \in V^{(\widehat{\mathbf{C}})}$.

POTENTIAL APPLICATIONS OF COVER SCHEMES, KRIPKE MODELS, AND FRAME-VALUED SET THEORY IN SPACETIME PHYSICS

In spacetime physics any set C of events—a *causal set*—is taken to be partially ordered by the relation $\leq$ of *possible causation*: for $p, q \in C$, $p \leq q$ means that q is in p*'s* future light cone. In her groundbreaking paper [5] Fotini Markopoulou proposes that the causal structure of spacetime itself be represented by "sets evolving over C" —that is, in essence, by the topos $\mathbf{Set}^{\mathrm{C}}$ of presheaves on C^{op}. To enable what she has done to be the more easily expressed within the framework presented here, we will reverse the causal ordering, that is, C will be replaced by C^{op}, and the latter written as *P*—which will, moreover, be required to be no more than a *preordered* set. Thus *P* is a set of events preordered by the relation $\leq$, where $p \leq q$ is intended to mean that *p* is in *q's* future light cone—that *q could* be the cause of *p*. In requiring that $\leq$ be no more than a preordering—in dropping, that is, the antisymmetry of $\leq$—we are, in physical terms, allowing for the possibility that the universe is of Gödelian type, containing closed timelike lines.

Specifically, then, we fix a preordered set $(P, \leq)$, which we shall call the *universal causal set*; its members will be called *events* and $p \leq q$ understood to mean that *p* is in *q's* causal *future,* or *q*'s future light cone, in short, that *p* is a possible *effect* of *q*. (Thus, for each event *p,* the set $p{\downarrow}$ *is p's* future light cone.) Markopoulou, in essence, suggests that viewing the universe "from the inside" amounts to placing oneself within the topos of presheaves or "evolving universe" $\mathbf{Set}^{P^{op}}$. Since, as we have already observed, $\mathbf{Set}^{P^{op}}$ is equivalent to the topos of sets in $V^{(\widehat{P})}$, Markopoulou's proposal may be effectively realized by working within $V^{(\widehat{P})}$. Let us do so, writing for simplicity *H* for $\widehat{P}$: we think of $V^{(H)}$ as an *evolving universe*, and describing what the universe looks like "from the inside" will then amount to reporting the view from $V^{(H)}$. Each sentence σ of the language of set theory will be construed as an *assertion* concerning the evolving universe $V^{(H)}$.

The fact that each truth value $[\![\sigma]\!]^H$ (which we shall normally abbreviate to $[\![\sigma]\!]$) is a sieve in *P*— that is, satisfies $p \in [\![\sigma]\!]$ and $q \leq p \rightarrow \quad q \in [\![\sigma]\!]$ may be understood as asserting that truth values in the evolving universe are "closed under potential effects", or "causally closed".

The forcing relation $\Vdash_P$ (which we will usually write simply as $\Vdash$) defined in the previous section now links events *p* and assertions σ: $p \Vdash \sigma$ will be taken to mean that σ *holds* as a result of (the occurrence of) event *p*, or that *p induces* the assertion σ to hold. The persistence of $\Vdash$ —i.e. the fact that, if $p \Vdash \sigma$ and $q \leq p$, then $q \Vdash \sigma$—amounts to the observation that, once an event *p* deduces an assertion to hold, that assertion continues to hold throughout *p's* causal future[5].

Define the set $K \in V^{(H)}$ by $\mathrm{dom}(K) = \{\hat{p} : p \in P\}$ and $K(\hat{p}) = p\downarrow$. Then, in $V^{(H)}$, *K* is a subset of $\widehat{P}$ and for $p \in P$, $[\![\hat{p} \in K]\!] = p\downarrow$. *K* is the counterpart in $V^{(H)}$ of the "evolving" set *Past* Markopoulou defines by $Past(p) = p{\downarrow}$, with insertions as transition maps. ($\widehat{P}$, incidentally, is the $V^{(H)}$- counterpart of the constant presheaf on *P* with value *P* which

---

5 It follows that assertions must be taken as being implicitly in the past tense: "such and such *was* the case".

Markopoulou calls *World*.) Accordingly the "causal past" of any "event" p is represented by the truth value in $V^{(H)}$ of the statement $\hat{p} \in K$. The fact that, for any $p, q \in P$ we have

$$q \Vdash_P \widehat{p} \in K \leftrightarrow q \leq p$$

may be construed as asserting that the events in the causal future of an event p are precisely those forcing (the canonical representative of) p to be a member of K. For this reason we shall call K the causal set in $V^{(H)}$.

If we identify each $p \in P$ with $p{\downarrow} \in H$, P may then be regarded as a subset of H so that, in $V^{(H)}$, $\widehat{P}$ is a subset of $\widehat{H}$. It is not hard to show that, in $V^{(H)}$, K generates the canonical prime filter Φ_H in $\widehat{H}$. Using the V-genericity of Φ_H, and the density of P in H, one can show that $[\![\sigma]\!] = [\![\exists p \in K. p \leq \widehat{[\![\sigma]\!]}]\!]$, so that, with moderate abuse of notation,

$$V^{(H)} \vDash [\sigma \leftrightarrow \exists p \in K.\, p \Vdash \sigma].$$

That is, in $V^{(H)}$, *a sentence holds precisely when it is forced to do so at some "causal past stage" in K.* This establishes the centrality of the causal set *K*—and, correspondingly, that of the "evolving" set *Past*— in determining the truth of sentences "from the inside", that is, inside the universe $V^{(H)}$.

Markopoulou also considers the complement of *Past*. In the present setting, this is the $V^{(H)}$-set $\neg K$—the *complement* of the causal set *K*—for which $[\![\widehat{p} \in \neg K]\!] = [\![p \notin K]\!] = \neg p{\downarrow}$. Markopoulou calls (*mutatis mutandis*) the events in $\neg p{\downarrow}$ those *beyond p's causal horizon*, in that no observer at p can ever receive "information" from any event in $\neg p{\downarrow}$. Since clearly we have

$$q \Vdash \widehat{p} \in \neg K \leftrightarrow q \in \neg p{\downarrow}, \qquad (\bigstar)$$

it follows that the events beyond the causal horizon of an event p are precisely those forcing (the canonical representative of) p to be a member of ¬K. In this sense ¬K reflects, or "measures" the causal structure of P.

In this connection it is natural to call $\neg\neg p{\downarrow} = \{q : \forall r \leq q \exists s \leq r. s \leq p]$ the *causal horizon* of *p*: it consists of those events *q* for which an observer placed at *p* could, in its future, receive information from any event in the future of an observer placed at *q*. Since

$$q \Vdash \widehat{p} \in \neg\neg K \leftrightarrow q \in \neg\neg p{\downarrow},$$

it follows that *the events within the causal horizon of an event are precisely those forcing (the canonical representative of) p to be a member of ¬¬K.*

It is easily shown that $\neg K$ is *empty* (i.e. $V^{(H)} \vDash \neg K = \varnothing$) if and only if P is *directed downwards* in the sense that for any $p, q \in P$ there is $r \in P$ for which $r \leq p$ and $r \leq q$; that is, if *the future light cones of any pair of events have nonempty intersection or "overlap"*. This holds in the case, considered by Markopoulou, of *discrete Newtonian time evolution*—in the present setting, the case in which P is the opposite $\mathbb{N}^{op}$ of the totally ordered set $\mathbb{N}$ of natural numbers. Here the corresponding complete Heyting algebra H is the family of all downward-closed sets of natural numbers. Interestingly, in this case, the causal set K *is neither finite nor actually infinite.*

To see this, first note that, for any natural number n, we have, $[\![\neg(\hat{n} \in \neg K)]\!] = \mathbb{N}$. It follows that $V^{(H)} \models \neg\neg\forall n \in \widehat{\mathbb{N}}\,.\, n \in K$. But, working in $V^{(H)}$, if $\forall n \in .\widehat{\mathbb{N}}\ n \in K$, then K is not finite, so if K is finite, then $\neg\forall n \in \widehat{\mathbb{N}}\,.\, n \in K$, and so $\neg\neg\forall n \in \widehat{\mathbb{N}}\,.\, n \in K$ implies the non-finiteness of K.

But, in $V^{(H)}$, K is not actually infinite. For (again working in $V^{(H)}$), if K were actually infinite (i.e., if there existed an injection of $\widehat{\mathbb{N}}$ into K), then the statement

$$\forall x \in K\ \exists y \in K.\ x > y$$

would also have to hold in $V^{(H)}$. But calculating that truth value gives:

$$\begin{aligned}
[\![\forall x \in K \exists y \in K. x > y]\!] &= \bigcap_{m \in \mathbb{N}^{op}} [m\downarrow \Rightarrow \bigcup_{n \in \mathbb{N}^{op}} n\downarrow \cap [\![\hat{m} > \hat{n}]\!]] \\
&= \bigcap_{m} [m\downarrow \Rightarrow \bigcup_{n<m} n\downarrow] \\
&= \bigcap_{m} [m\downarrow \Rightarrow (m+1)\downarrow] \\
&= \bigcap_{m} (m+1)\downarrow \\
&= \varnothing
\end{aligned}$$

So $\forall x \in K\ \exists y \in K.\ x > y$ is false in $V^{(H)}$ and therefore K is not actually infinite.

In other words, in evolving Newtonian spacetime, the set K representing past time is potentially, but not actually infinite: this is, in essence, what Kant asserted of time.

In order to formulate an observable causal *quantum theory* Markopoulou considers the possibility of introducing a *causally evolving algebra of observables.* This amounts to specifying a presheaf of C^*-algebras on P, which, in the present framework, corresponds to specifying a set $\mathtt{A}$ in $V^{(H)}$ satisfying

$$V^{(H)} \models \text{A is a C*-algebra.}$$

The "internal" C^*-algebra $\mathtt{A}$ is then subject to the intuitionistic internal logic of $V^{(H)}$: *any* theorem concerning C^*-algebras—provided only that it be constructively proved—automatically applies to $\mathtt{A}$. Reasoning with $\mathtt{A}$ is more direct and simpler than reasoning with A.

This same procedure of "internalization" can be performed with any causally evolving object: each such object of type $\mathtt{T}$ corresponds to a set S in $V^{(H)}$ satisfying

$$V^{(H)} \models \text{ S is of type T.}$$

Internalization may also be applied in the case of the presheaves *Antichains* and *Graphs* considered by Markopoulou. Here, for each event *p, Antichains*(*p*) consists of all sets of causally unrelated events in *Past*(*p*), while *Graphs*(*p*) is the set of all graphs supported by elements of *Antichains*(*p*). In the present framework *Antichains* is represented by the $V^{(H)}$ –set $Anti = \{ X \subseteq \widehat{P} : X \text{ is an antichain}\}$ and *Graphs* by the $V^{(H)}$ –set $Grph = \{G\text{: } \exists X \in A \,.G \text{ is a graph supported by } A\}$. Again, both *Anti* and *Grph* can be readily handled using the internal intuitionistic logic of $V^{(H)}$.

Finally let us examine the role of cover schemes on causal sets. Suppose we are given a cover scheme **C** on the universal causal set *P*. Each **C**-cover of an event *p* may be thought of as a "sampling" of the events in *p*'s causal future, a "survey" of *p*'s potential effects—in a word, a *survey of p*. Using this language the defining condition **(Cov)** for cover schemes laid down in section I becomes: *for any survey S of a given event p, and any event q which is a possible effect of p, there exists a survey of q each event in which is the possible effect of some event in S.*

As we have seen, cover schemes may be used to force certain conditions to prevail in the associated models. Let us consider, for example, the cover scheme **Den** in *P*. We know that the associated frame $\widehat{\mathbf{Den}}$ is a Boolean algebra—let us denote it by *B*. The corresponding causal set K_B in $V^{(B)}$ then has the property

$$[\![\widehat{p} \in K_B]\!] = \neg\neg p \downarrow;$$

so that,

$$q \Vdash_B \widehat{p} \in K_B \leftrightarrow q \in \neg\neg p\downarrow$$

$$\leftrightarrow \text{ q is in p's causal horizon.}$$

Comparing this with (★) above, we see that moving to the universe $V^{(B)}$—"Booleanizing" it, so to speak—*amounts to replacing causal futures by causal horizons.* When *P* is linearly ordered, as for example in the case of Newtonian time, the causal horizon of any event coincides with the whole of *P*, *B* is the two-element Boolean algebra **2,** so that $V^{(B)}$ is just the universe *V* of "static" sets. In this case, then, the effect of "Booleanization" is to *render the universe timeless.*

The universes associated with the cover schemes $\mathbf{C}^A$ and $\mathbf{C}_A$ seem also to have a rather natural physical meaning. Consider, for instance the case in which *A* is the sieve $p\downarrow$—the causal future of *p*. In the associated universe $V^{(\widehat{\mathbf{C}^A})}$ the corresponding causal set K^A satisfies

$$[\![\widehat{q} \in K^A]\!] = \textit{least } \mathbf{C}^{\mathbf{A}}\text{-closed sieve containing q}$$

so that, in particular

$$[\![\widehat{p} \in K^A]\!] = \textit{least } \mathbf{C}^{\mathbf{A}}\text{-closed sieve containing p}$$

$$= \text{P}.$$

This means that, for every event q,

$$q \Vdash_{\widehat{\mathbf{C}^{\mathbf{A}}}} \widehat{p} \in K^A.$$

Comparing this with (★), we see that in $V^{(\widehat{\mathbf{C}^A})}$ that every event has been "forced" into p's causal future: in short, that p now marks the "beginning" of the universe as viewed from inside $V^{(\widehat{\mathbf{C}^A})}$.

Similarly, we find that the causal set K_A in the universe $V^{(\widehat{\mathbf{C}_A})}$ has the property

$$q \leq p \rightarrow \forall r[r \Vdash_{\widehat{\mathbf{C}_{\mathbf{A}}}} \widehat{q} \in \neg K_A].$$

a comparison with (★) above reveals that, in $V^{(\widehat{\mathbf{C}^A})}$, every event—including p itself—has been placed beyond p's causal horizon. In effect, the event p has been obliterated, effaced from the universe—like the extraordinary events in H.G. Wells' *The Man Who Could Work Miracles,* the event p never occurred!

As a final possibility consider the universe $V^{(\widetilde{\widetilde{P}})}$ associated with the free lower semilattice $\widetilde{P}$ generated by P. In this case the elements of $\widetilde{P}$ are finite sets of events, preordered by the relation $\sqsubseteq$: for F, $G \in \widetilde{P}$, $F \sqsubseteq G$ iff every event in G is in the causal past of an event in F. The empty set of events is the top element of $\widetilde{P}$. The causal set $\widetilde{K}$ in $V^{(\widetilde{\widetilde{P}})}$ has the property that its complement $\neg\widetilde{K}$ is empty (so that, in this universe, the light cones of any pair of "events" overlap) and $\widehat{\varnothing}$ is an initial event in the sense that $F \Vdash_{\widetilde{P}} \widehat{\varnothing} \in \widetilde{K}$ for every "event" F. In this case passage to the new universe $V^{(\widetilde{\widetilde{P}})}$ preserves the original causal relations in the sense that

$$\{q\} \Vdash_{\widetilde{P}} \widehat{\{p\}} \in \widetilde{K} \leftrightarrow q \Vdash_P \widehat{p} \in K.$$

In other words, in passing to the new universe the initial event $\widehat{\varnothing}$ and the new light cone overlaps have been "freely adjoined" to the original universe.

References

[1] Bell, J. L. Boolean-Valued Models and Independence Proofs in Set Theory. Oxford University Press, 1977.

[2] Grayson, R. J. Heyting-valued models for intuitionistic set theory. In: Applications of Sheaves. Springer Lecture Notes in Mathematics **753**, 1979.

[3] Johnstone, P.T. Stone Spaces. Cambridge University Press, 1982.

[4] Mac Lane, S. and Moerdijk, I. Sheaves in Geometry and Logic. Springer-Verlag, 1992.

[5] Markopoulou, F. The internal description of a causal set: What the universe looks like from the inside. arXiv:gr-qc/98ii053 v2 18 Nov 1999.

[6] Rosenthal, K. I. Quantales and their Applications. Longman, 1990.

In: Horizons in World Physics, Volume 248
Editor: A. Reimer, pp. 75-110
ISBN 1-59454-322-4

Chapter 4

A GLOBAL AND NON-ENTROPIC APPROACH TO THE PROBLEM OF THE ARROW OF TIME

Mario Castagnino* and Olimpia Lombardi#
*CONICET - Instituto de Astronomía y Física del Espacio, Argentina
#CONICET - Universidad Nacional de Quimes, Argentina

ABSTRACT

Since the nineteenth century, the problem of the arrow of time has been traditionally analyzed in terms of entropy by relating the direction past-to-future to the direction of the entropy increase in isolated systems. In this paper, we reject this traditional perspective and argue for a global and non-entropic approach to the problem, according to which the arrow of time can be defined in terms of the geometrical properties of spacetime. In particular, we show how the global non-entropic arrow can be transferred to the local level, where it takes the form of a non-spacelike local energy flow that provides the criterion for breaking the symmetry resulting from time-reversal invariant local laws.

I. INTRODUCTION

The problem of the arrow of time was born in the nineteenth century in the context of the first discussions about the meaning and the mechanical interpretation of the second law of thermodynamics. Since those days, the problem has been traditionally analyzed in terms of entropy, by defining the future as the direction of time in which the entropy of isolated systems increases. Nevertheless, the debates about this subject have continued up to now: at present the problem of the arrow of time is still one of the most controversial questions in the foundations of physics.

In this paper we will move away from the mainstream views about the arrow of time by addressing the problem from a global and non-entropic perspective, according to which the arrow of time can be defined in terms of the geometrical properties of spacetime, independently of any entropic consideration. In this sense, we will analyze under what conditions it is possible to define a global and non-entropic arrow, and how such a global arrow can be transferred to the local level as a non-spacelike energy flow that provides the criterion for breaking the symmetry resulting from the time-reversal invariant laws of local physics.

Although we have devoted several papers to this subject ([1], [2]), we are aware of the fact that our global and non-entropic approach may sound strange to those who believe in an inescapable link between the arrow of time and thermodynamic notions. For this reason, we will devote the following two sections of the paper to clarify concepts and to justify our perspective: in Section II the concepts of arrow of time and of irreversibility are distinguished and the corresponding problems are clearly disentangled; in Section III the global and non-entropic character of our approach is justified by contrast with traditional views. The rest of the paper is organized as follows. In Section IV we consider the necessary conditions for defining a global and non-entropic arrow of time. Section V is devoted to introduce the definition of the global and non-entropic arrow of time in geometrical terms, and to prove the generic nature of such an arrow. In Section VI the arguments developed in the two previous sections are applied to the case of Friedmann-Lemâitre-Robertson-Walker (FLRW) models. Section VII is devoted to explain under what conditions the geometrical arrow can be translated into a global energy flow defined in terms of the energy-momentum tensor, and how that global energy flow can be transferred to local contexts as a local energy flow. Finally, In Section VIII we will show how the local energy flow supplies the necessary criterion for distinguishing between the time-symmetric structures resulting from local physical theories.

II. Disentangling Problems: Irreversibility and Arrow of Time

When the problem of the arrow of time is addressed, the main obstacle to be faced is conceptual confusion: the lack of consensus is primarily due to the fact that different concepts are identified and different problems are subsumed under the same label. In particular, the problem of irreversibility and the problem of the arrow of time are usually identified, as if irreversibility were the clue for understanding the origin and the nature of the arrow of time. In the present section we will show that, when the concepts involved in the debate are precisely clarified, both problems become evidently different. In particular, we will see that the question of irreversibility presupposes the previous answer to the problem of the arrow of time.

The Problem of Irreversibility

In general, the concepts of irreversibility and of time-reversal invariance are invoked in the discussions about the arrow of time, but usually with no elucidation of their precise meanings; this fact leads to confusions that contaminate many interesting debates. For this reason, we will begin by providing some necessary definitions.

Definition 1: A dynamical equation is *time-reversal invariant* if it is invariant under the application of the time-reversal operator $\mathbf{T}$, which performs the transformation $t \to -t$ and reverses all the dynamical variables whose definitions in function of t

are non-invariant under the transformation $t \to -t$. As a result, for each solution $e(t)$, $\mathbf{T}e(t)$ is also a solution.

When the dynamical equation can be expressed as $e(t) = U_t\, e(0)$, where U_t is a unitary evolution operator such that $U_{-t} = \mathbf{T}^{-1} U_t \mathbf{T}$, the equation is time-reversal invariant when:

$$\mathbf{T}e(0) = U_t\, \mathbf{T} U_t\, e(0) \tag{1}$$

It can be proved that, in this case, the time-reversal invariance condition amounts to the fact that the set of the operators U_t forms a group in such a way that $U_t U_{-t} = I$.

Definition 2: A solution of a dynamical equation is *reversible* if it has no attractors in finite or infinite points of the phase space.

This definition can be easily generalized to statistical and quantum cases: an evolution is reversible if it has no final (or initial) equilibrium state, namely, a "point of no return".

It is quite clear that both concepts are different to the extent that they apply to different mathematical entities: time-reversal invariance is a property of dynamical equations (laws) and, *a fortiori*, of the set of its solutions (evolutions); reversibility is a property of a single solution of a dynamical equation. Furthermore, they are not even correlated. In fact, both properties can combine in the four possible alternatives:

- **Time-reversal invariance and reversibility.** Let us consider the harmonic oscillator with Hamiltonian:

$$H = \frac{1}{2} p^2 + \frac{1}{2} K^2 q^2 \tag{2}$$

The equation is time-reversal invariant, namely, it is invariant under the transformation $q \to q,\ p \to -p$. As a result, the set of trajectories is symmetric with respect to the q-axis. Since each trajectory is a closed ellipse, it is reversible.

- **Time-reversal invariance and irreversibility.** Let us consider the pendulum with Hamiltonian:

$$H = \frac{1}{2} p_\theta^2 + \frac{K^2}{2} \cos\theta \tag{3}$$

Again the equation is time-reversal invariant, and the set of solutions is symmetric with respect the θ-axis. The trajectories within the separatrices are reversible since they are closed curves. But the trajectory corresponding to the upper (lower) separatrix is irreversible since it tends to $\theta = \frac{\pi}{2}, p_\theta = 0$ $(\theta = -\frac{\pi}{2}, p_\theta = 0)$ when $t \to \infty$ $(t \longrightarrow -\infty)$.

- **Non time-reversal invariance and reversibility.** Let us consider the modified oscillator with Hamiltonian:

$$H = \frac{1}{2} p^2 + \frac{1}{2} K(p)^2 q^2 \tag{4}$$

where $K(p) = K^{+} = const$ when $p \geq 0$, and $K(p) = K^{-} = const$ when $p < 0$. If $K^{+} \neq K^{-}$, the equation is not time-reversal invariant: the set of solutions is clearly asymmetric with respect to the q-axis. Nevertheless, each trajectory is closed and, therefore, reversible.

- **Non time-reversal invariance and irreversibility.** Let us consider a damped oscillator represented by the following equation:

$$\ddot{q} = -K^2 q - A^2 \dot{q} \tag{5}$$

The equation is non time-reversal invariant. On the other hand, the origin is an attractor and the trajectories are spirals: they are trapped by the origin when $t \to \infty$. Therefore, each trajectory is irreversible.

When both concepts are elucidated in this way, *the problem of irreversibility* can be clearly stated: *how to explain irreversible evolutions in terms of time-reversal invariant laws*. Let us note that this definition of the problem includes the usual characterization in terms of thermodynamic concepts, since entropy increase is a feature of irreversible thermodynamic evolutions that should be explained in terms of the time-reversal invariant laws of mechanics. But our definition is more general than the usual characterization since it also includes other interesting cases of irreversibility, like those studied in the so-called irreversible quantum mechanics.

Once it is recognized that irreversibility and time-reversal invariance apply to different mathematical entities, it turns out to be clear that there is no conceptual puzzle in the problem of irreversibility: nothing prevents a time-reversal invariant equation from having irreversible solutions. Of course, although the conceptual answer is simple, a great deal of theoretical work is needed for obtaining irreversible evolutions from an underlying time-reversal invariant dynamics; usually it is necessary to appeal to formal tools as coarse-grainings, traces, restricted algebras, etc. This was the problem faced by the founding fathers of statistical mechanics when they sought to describe the irreversible evolutions of thermodynamics by means of the time-reversal invariant laws of classical mechanics. At present, many efforts are directed to explain the irreversible behavior of quantum systems in terms of the time-reversal invariant quantum theory (see, for instance, [3], [4], [5]). But the point to stress here is that the question about the arrow of time does not need to be invoked for addressing the problem of irreversibility. In fact, when we talk about entropy increasing processes, we are presupposing an entropy increase *towards the future*: or when we consider a process going from non-equilibrium to equilibrium, we implicitly locate equilibrium *in the future*. In general, any evolution that tends to an attractor is conceiving as approaching it towards the future. This means that the distinction between past and future is *usually presupposed* in the traditional treatments of the problem of irreversibility[1]. But this is not a shortcoming of those treatments, since their aim is to explain irreversibility and not to seek a physical distinction between the two directions of time. On the contrary, such a distinction becomes the central point when the problem of the arrow of time is the question at issue.

[1] When the distinction between past and future is not presupposed, time-reversal invariant dynamical laws lead to the "time-symmetric twins" problem, which will be discussed in Section VIII.

The Problem of the Arrow of Time

The problem of the arrow of time owes its origin to the intuitive asymmetry between past and future. We experience the time order of the world as "directed": if two events are not simultaneous, one of them is earlier than the other. Moreover, we view our access to past and future quite differently: we remember the past and predict the future. The ultimate metaphysical nature of time has been one of the traditional interests of philosophy since its birth. There seems to be something essentially evasive in our experience of time and its "flow" from past to future through the present. Some authors attempt to ground the intuitive asymmetry between past and future on the increase of entropy in isolated systems. An argument that has been presented in this context concerns to traces: there are many traces of the past but none of the future (see the argument of the "footprint in the sand" in [6] and [7], and the criticisms in [8]). In other cases, the experience of the "directedness" of time is attempted to be explained in terms of biology: the functioning of the biological mechanisms which underlie our perception of time order and our sense of time depend of the entropic behavior of isolated systems. But there seems to be little or no evidence in support to this claim; on the contrary, scientific research suggests that the perception of time order is more subtle and complex than that position assumes.

Here we will not discuss the questions related with the time-asymmetry of our experience or perception of time. As in the case of irreversibility, we will address the problem of the arrow of time within the limits of physics. In this context, the problem of the arrow of time arises when we seek a *physical correlate* of this intuitive asymmetry between past and future: do physical theories pick out a preferred direction of time?

The main difficulty to be encountered in answering this question relies on our anthropocentric perspective: the difference between past and future is so deeply rooted in our language and our thoughts that it is very difficult to shake off those temporally asymmetric assumptions. In fact, traditional discussions around the problem of the arrow of time are usually subsumed under the label "the problem of the direction of time", as if we could find an exclusively physical criterion for singling out *the* direction of time, identified with what we call "the future". But there is nothing in physical evolution laws that distinguishes, in a non-arbitrary way, between past and future as we conceive them in our ordinary language and our everyday life. It might be objected that physics implicitly assumes this distinction with the use of temporally asymmetric expressions, like "future light semicone", "initial conditions", "increasing time", and so on. However this is not the case, and the reason relies on the distinction between "conventional" and "substantial".

> **Definition 3:** Two objects are *formally identical* when there is a permutation that interchanges the objects but does not change the properties of the system to which they belong.

In physics it is usual to work with formally identical objects: the two semicones of a light cone, the two spin senses, etc.

> **Definition 4:** We will say that we establish a *conventional* difference between two objects when we call two formally identical objects with two different names.

This is the case when we assign different signs to the two spin senses, or different names to the two light semicones, etc.

> **Definition 5:** We will say that the difference between two objects is *substantial* when we assign different names to two objects that are not formally identical (see

[9], [10]). In this case, although the particular names we choose are conventional, the difference is substantial.

For instance, the difference between the two poles of the theoretical model of a magnet is conventional since both poles are formally identical; on the contrary, the difference between the two poles of the Earth is substantial because in the North Pole there is an ocean and in the South Pole there is a continent (and the difference between ocean and continent remains substantial even if we conventionally change the names of the poles).

Once this point is accepted, it turns out to be clear that physics uses the labels "past" and "future" in a conventional way. Therefore, the problem cannot yet be posed in terms of singling out the future direction of time: the problem of the arrow of time becomes the problem of finding a *substantial difference between the two temporal directions*. But if this is our central question, we cannot project our independent intuitions about past and future for solving it without begging the question. When we want to address the problem of the arrow of time from a perspective purged of our temporal intuitions, we must avoid the conclusions derived from subtly presupposing time-asymmetric notions. As Huw Price [11] claims, it is necessary to stand at a point outside of time, and thence to regard reality in atemporal terms: this is his "*view from nowhen*". This atemporal standpoint prevents us from using temporally asymmetric expressions in a non-conventional way: the assumption about the difference between past and future is not yet legitimate in the context of the problem of the arrow of time.

But then, what does "the arrow of time" mean when we accept this constraint? Of course, the traditional expression coined by Eddington has only a metaphorical sense: its meaning must be understood by analogy. We recognize the difference between the head and the tail of an arrow on the basis of its geometrical properties; therefore, we can substantially distinguish between both directions, head-to-tail and tail-to-head, *independently of our particular perspective.* Analogously, we will conceive *the problem of the arrow of time* in terms of *the possibility of establishing a substantial distinction between the two directions of time on the basis of exclusively physical arguments*.

III. A Global and Non-Entropic Approach

Why Global? The Traditional Local Approach

The traditional local approach owes its origin to the attempts to reduce thermodynamics to statistical mechanics: in this context, the usual answer to the problem of the arrow of time consists in defining the future as the direction of time in which entropy increases. However, already in 1912 Paul and Tatiana Ehrenfest [12] noted that, when entropy is defined in statistical terms on the underlying classical dynamics, if the entropy of a closed system increases towards the future, such increase is matched by a similar one in the past of the system. In other words, if we trace the dynamical evolution of a non-equilibrium system at the initial time back into the past, we will obtain states that are closer to the equilibrium state. Gibbs' answer to the Ehrenfests' challenge was based on the assumption that "*while probabilities of subsequent events may be often determined from probabilities of prior events, it is rarely the case that probabilities of prior events can be determined from those of subsequent events*" (quoted in [8]). But this answer clearly violates the "nowhen" standpoint, since probabilities are blind to temporal direction when they evolve according to time-reversal equations: probabilities assignments based on previous events presuppose a sort of asymmetric causality which, in turn, takes for granted the existence of asymmetric temporal relationships between events. Therefore, any appeal to the distinction between prior and subsequent events commits a *petitio principii* by presupposing the arrow of time from the very beginning.

The point can also be posed in different terms. Let us assume that we have solved the irreversibility problem; so we have the description of all the irreversible evolutions, say, decaying processes, of the universe. However, since we have not yet established a substantial difference between both directions of time, we have no way to decide towards which temporal direction *each* decay proceeds. Of course, we would obtain the arrow of time if we could coordinate the processes in such a way that all of them parallely decay towards the same temporal direction. But this is precisely what local physics cannot offer: only by means of global considerations all the decaying processes can be coordinated. This means that the global arrow of time plays the role of the background scenario where we can meaningfully speak of the temporal direction of irreversible processes, and this scenario cannot be built up by means of local theories that only describe phenomena confined in small regions of spacetime.

Once it is accepted that the problem of the arrow of time cannot be addressed from a local perspective, general relativity comes into play. As it is well known, in this theory the only basic properties are the geometrical properties of spacetime, embodied in the metric tensor, and the distribution of matter-energy throughout the spacetime, embodied in the energy-momentum tensor. But since both properties are related by the Einstein's field equations, the substantial difference between the two directions of time can be conceived as a geometrical property of spacetime.

Why Non-Entropic? The Traditional Entropic Approach

When, in the late nineteenth century, Boltzmann developed the probabilistic version of his theory in response to the objections raised by Loschmidt and Zermelo (for historical details, see [13]), he had to face a new challenge: how to explain the highly improbable current state of our world. In order to answer this question, Boltzmann offered the first global approach to the problem: "*The universe, or at least a big part of it around us, considered as a mechanical system, began in a very improbable state and it is now also in a very improbable state. Then if we take a smaller system of bodies, and we isolate it instantaneously from the rest of the world, in principle this system will be in an improbable state and, during the period of isolation, it will evolve towards more probable states*" [14]. Since this seminal work, many authors have related the temporal direction past-to-future to the gradient of the entropy function of the universe: it has been usually assumed that the only way of distinguishing between the two directions of time is by means of the second law of thermodynamics. Even those who admit the existence of many arrows of time tend to suppose that thermodynamics supplies the fundamental criterion for time-asymmetry: all the other arrows should be derived, in some sense, from the entropic arrow.

One of the most philosophically influential works on the problem of the arrow of time was *The Direction of Time* of Hans Reichenbach [15], where the future is defined as the direction of the entropy increase of the majority of the *branch systems* of the universe, that is, systems which become isolated from the main system during certain period. Paul Davies also appeals to Reichenbach's notion of branch systems [16], and considers that those systems emerge as the result of a chain or hierarchy of branchings which expand out into wider and wider regions of the universe: "*the origin of the arrow of time always refer back to the cosmological initial conditions. There exists an arrow of time only because the universe originates in a less-than-maximum entropy state*" [17]. In a similar sense, Feynman claims that "*For some reason, the universe at one time had a very low entropy for its energy content, and since then entropy has increased. So that is the way towards future. That is the origin of all irreversibility*" [18]. Even if these authors admit the need of global arguments for solving the problem of the arrow of time, they agree in considering that it must be addressed in terms of entropy.

The global entropic approach rests on two assumptions: that it is possible to define entropy for a complete cross-section of the universe, and that there is an only time for the universe as a whole. However, both assumptions involve difficulties. In the first place, the definition of entropy in cosmology is still a very controversial issue: there is no consensus regarding how to define a global entropy for the universe. In fact, it is usual to work only with the entropy associated with matter and radiation because there is not yet a clear idea about how to define the entropy due to the gravitational field. In the second place, when general relativity comes into play, time cannot be conceived as a background parameter which, as in pre-relativistic physics, is used to mark the evolution of the system. Therefore, the problem of the arrow of time cannot legitimately be posed, from the beginning, in terms of the entropy gradient of the universe computed on a background parameter of evolution.

Nevertheless, these points are not the main difficulty: there is a conceptual argument for abandoning the traditional entropic approach. As it is well known, a given value of entropy is compatible with many configurations of a system: entropy is a phenomenological property. The question is whether there is a more fundamental property of the universe which allows us to distinguish between both temporal directions. On the other hand, if the arrow of time reflects a substantial difference between both directions of time, it is reasonable to consider it as an intrinsic property of time, or better, of spacetime, and not as a secondary feature depending on a phenomenological property. For these reasons we will follow Earman's "*Time Direction Heresy*" [19], according to which the arrow of time is an intrinsic, geometrical property of spacetime embodied in $g_{\mu\nu}(x)$, which does not need to be reduced to non-temporal features. In other words, the geometrical approach to the problem of the arrow of time has conceptual priority over the entropic approach, since the geometrical properties of the universe are more basic than its thermodynamic properties.

Why Generic?

As we have seen, the global entropic approach explains the arrow of time in terms of the gradient of the entropy function of the universe: the future direction of time is, by definition, the direction in which the entropy of the universe increases. As a consequence, this position has to posit a low-entropy initial state from which entropy increases (see quotes of Feynman and Davies above). Then, the problem of the arrow of time is pushed back to the question of why the initial state of the universe has low entropy. But a low-entropy initial state is extraordinarily improbable in the collection of all possible initial states. Therefore, the global entropic approach is committed to supply an answer to the problem of explaining such an improbable initial condition.

According to Davies [17], the expansion of the early universe during the first few minutes caused it to depart from equilibrium; but this fact requires to suppose that "*the gravitational field at the time of the big bang was in a much-less-than-maximum entropy state*". Penrose [20], in turn, suggests that the early universe begins with a high degree of spatial homogeneity and isotropy; but this uniform distribution is actually a low-entropy state when gravity is taken into account, since the purely attractive force of gravity is such that a uniform distribution is stupendously improbable. In either case, the global entropic approach has to face the problem of explaining an extraordinarily improbable initial condition such as the low-entropy or the near-flatness of the spacetime.

In our global and non-entropic approach there are not improbable conditions which require an explanatory account. On the contrary, we will prove that, if the spacetime has some minimal topological properties like time-orientability and global time (which, anyway, are also necessary conditions for the existence of the arrow of time in the global entropic approach), the arrow of time is a *generic* feature of the universe. This means that, in the collection of all possible spacetimes, the universes endowed with a global and non-entropic arrow of time are overwhelmingly probable: the non-existence of the arrow of time is what requires an extraordinarily fine-tuning of all the variables of the universe.

IV. Conditions for a Global and Non-Entropic Arrow of Time

Time-Orientability

In a Minkowski spacetime, it is always possible to define the class of all the future light semicones and the class of all the past light semicones (where the labels "future" and "past" are conventional). In general relativity the metric can always be approximated, in small regions of spacetime, to the Minkowski form. However, on the large scale, we do not expect the manifold to be flat because gravity can no longer be neglected. Many different topologies are consistent with Einstein's field equations. In particular, the possibility arises of spacetime being curved along the spatial dimension in such a way that, e.g., the spacelike sections of the universe become the three-dimensional analogue of a Moebius band; in this case it is said that the spacetime is non time-orientable.

Definition 6: A *space-time* is an ordered pair (M,g), where M is a four-dimensional differentiable manifold, and g is a Lorentzian metric for M such that $\nabla g = 0$. For any point $x \in M$, M_x is the tangent space to M at x. The set $C(x)$ of all vectors $\mathbf{v} \in M_x$ such that $g(\mathbf{v},\mathbf{v}) = 0$ is the *light cone* at x. A vector $\mathbf{v} \in M_x$ is *timelike* (*null*, *spacelike*) if $\mathbf{v}$ lies inside (on, outside) $C(x)$.

Definition 7: A spacetime (M, g) is *time-orientable* if there exists a continuous non-vanishing vector field $\gamma^{\mu}(x)$ on M which is everywhere non-spacelike.

By means of this field, the set of all the semicones of the manifold can be split into two equivalence classes, C_+ and C_-: the semicones of C_+ contain the vectors of the field and the semicones of C_- do not contain them. On the other hand, in a non time-orientable spacetime it is possible to transform a future-pointing timelike vector into a past-pointing timelike vector by means of a continuous transport that always keeps non-vanishing timelike vectors timelike; therefore, the distinction between future semicones and past semicones is not definable in an univocal way on a global level. This means that the time-orientability of spacetime is a precondition for defining a global arrow of time since, if spacetime is non time-orientable, it is not possible to distinguish between the two temporal directions for the universe as a whole.

However, not all accept the relevance of time-orientability to the problem of the arrow of time. For instance, Matthews [21] admits that we must rule out non time-orientable spacetimes if we want to define the arrow of time for the universe as a whole. But, for him, the problem can be posed in exclusively local terms; so, considerations of general relativity are of little or no importance. In particular, he claims that a spacetime may have a regional but not a global arrow of time and, in this case, time-orientability is not a necessary condition. In this sense, Matthews agrees with Reichenbach [15], who asserts that "*only certain sections of time have directions and these directions are not the same*".

Although these arguments may seem reasonable, even from this local approach (which we have rejected in the previous section for different reasons) time-orientability cannot be avoided. Let us suppose that there were a local non time-reversal invariant law L, which defines regional arrows of time that disagree when compared by means of continuous transport of timelike vectors. The trajectory of the transport will pass through a frontier point between both regions: in a region around this point the arrow of time will be not unequivocally defined, and this amounts to a breakdown of the validity of L in such a point. But this fact contradicts the methodological *principle of universality*, unquestioningly accepted in contemporary cosmology, according to which the laws of physics are valid in all points of the spacetime. The strategy to escape this conclusion would consist in refusing to assign any meaning to the continuous transport of timelike vectors. This strategy would only be acceptable if the two regions with different arrows were physically isolated: this amounts to the disconnectedness of the spacetime. But this fact would contradict another methodological principle of cosmology, that is, the *principle of uniqueness*, according to which there is only one universe and completely disconnected spacetimes are not allowed[2]. These arguments show that the possibility of arrows of time pointing to opposite directions in different regions of the spacetime not only requires a local definition of these arrows, but also implies to ignore basic considerations coming from contemporary cosmology.

---

[2] Although there are quantum cosmologies exhibiting disconnected spacetimes, such models only play an explanatory role since they are not testably in principle. Anyway, even if disconnected spacetimes were allowed, each connected region could be considered as a universe by itself, where continuous transport must be valid. This fact is relevant since we are interested in explaining the arrow of time of our own connected universe. For a different opinion, see [22] and our criticism in [2]

Global Time

As it is well known, general relativity replaces the older conception of space-through-time by the concept of spacetime, where space and time are inextricably interwinned. In one sense, time becomes a dimension of a four-dimensional manifold. But when the time measured by a physical clock is considered, each particle of the universe has its own *proper time*, that is, the time registered by the clock carried by the particle. Since the curved spacetime of general relativity can be considered locally flat, it is possible to synchronize the clocks fixed to particles whose parallel trajectories are confined in a small region of spacetime. But, in general, the synchronization of the clocks fixed to all the particles of the universe is not possible. Only in certain particular cases it is possible to talk of *the* time of the universe.

The issue can also be posed in geometrical terms. A spacetime may be such that it is not possible to partition the set of all events into equivalent classes such that: (i) each one of them is on a spacelike hypersurface, and (ii) the hypersurfaces can be ordered in time. This is the case when there are closed or almost closed timelike curves or, even without such kind of timelike curves, when it is impossible to find a smooth function that attaches to each event a real number, the time of the event, such that the number assigned to e_1 is less than that assigned to e_2 whenever there is a causal signal propagable from e_1 to e_2. In such cases the spacetime is not globally splittable into spaces-at-a-time, that is, into spacelike hypersurfaces each one of which contains all the events simultaneous with each other.

There is a hierarchy of conditions which, applied to a time-orientable spacetime, avoid the "anomalous" temporal features just described (see [23]). Taking the spacetime (M,g) to be time-orientable and using the words "future" and "past" in a conventional way, one can define:

> **Definition 8**: For $x \in M$, the *chronological future* (*past*) of x, $I^+(x)$ ($I^-(x)$), is the set of all $y \in M$ such that y can be reached from x by a future directed (past directed) timelike curve.
>
> **Definition 9**: For $x \in M$, the *causal future* (*past*) of x, $J^+(x)$ ($J^-(x)$), is the set of all $y \in M$ such that y can be reached from x by a future directed (past directed) non-spacelike curve.

The first condition of the hierarchy is the *chronology condition*, which holds on (M,g) iff $\forall x \in M$, $I^+(x) \cap I^-(x) = \varnothing$; this first condition rules out closed timelike curves. The second condition, stronger than the previous one, is the *causality condition*, which holds on (M,g) iff $\forall x \in M$, $J^+(x) \cap J^-(x) = \varnothing$; this condition rules out closed non-spacelike curves. If we also want to exclude cases where there are non-spacelike curves which return arbitrarily close to their point of origin, we must strengthen the requirement imposed on the spacetime: the *strong causality condition* holds on (M,g) iff $\forall x \in M$, every neighborhood of x contains a neighborhood of x which no non-spacelike curve intersects more than once. We may also want to exclude situations with spacelike curves that pass arbitrarily close to another spacelike curve that passes arbitrarily close to the origin of the first curve. As Hawking and Ellis point out [23], there is an infinite hierarchy of higher degree causality conditions depending on the number and order of the limiting processes involved. Even if the strong causality condition rules out closed and almost closed non-spacelike curves, it does not exclude the possibility that the slightest variation of the metric of the spacetime leads again to such temporal "pathologies". In order to avoid this possibility, the spacetime must have some form of stability, defined as a property of "nearby" spacetimes. The *stable causality condition* holds on (M,g) iff the strong causality condition holds on (M,g), and for every metric h which is sufficiently close to g, (M,h) has no closed timelike curves[3]. It can be proved that the stable causality condition holds on (M,g) iff (M,g) possesses a global time function:

Definition 10: A *global time function* on the spacetime (M,g) is a function $t: M \rightarrow R$ whose gradient is everywhere timelike.

The value of the global time function increases along every future directed non-spacelike curve. The existence of such a function guarantees that the spacetime is globally splittable into spaces-at-a-time, that is, into hypersurfaces of simultaneity ($t = const$) which define a *foliation* of the spacetime (see [24]).

Of course, the existence of a global time imposes a significant topological limitation on the spacetime. But with no global time, there is not a single time of the universe and, therefore, it is nonsensical to speak of the two directions of time for the universe as a whole. Therefore, the possibility of defining a global time is a precondition for meaningfully speaking of a global arrow of time.

This fact supplies an additional argument against the global entropic approach, which takes for granted the possibility of defining the entropy function of the universe. This amounts to the assumption that the spacetime can be partitioned into spacelike hypersurfaces on which the entropy of the universe can be defined (that is, the spacetime has a global time function). When the possibility of spacetimes with no global time is recognized, it is difficult to deny the conceptual priority of the geometrical structure of spacetime over its entropic features in the context of our problem.

V. The Definition of the Global and Non-Entropic Arrow of Time

Time-Asymmetry

It is quite clear that time-orientability and existence of global time are merely necessary conditions for defining the global arrow of time since they assume only the globally consistent oppositeness of the two directions of a single time for the universe as a whole. But time-orientability and existence of global time do not provide a physical, substantial criterion for distinguishing between the two directions of time. It is usually claimed that we would only have a straightforward method for defining the arrow of time if physics included some non

[3] Of course, in order to give a precise meaning to "close", it is necessary to define a topology on the set of all spacetimes.

time-reversal invariant law. The puzzle seems to arise because the fundamental laws of physics are invariant under time-reversal[4]. Nevertheless, this usual opinion can be objected when the concept of time-symmetry is clearly elucidated.

As we have seen, time-reversal invariance is a property that a law possesses when it is invariant under time-reversal. But this concept must be distinguished from the concept of time-symmetry.

Definition 11: A solution $e(t)$ of a dynamical equation is *time-symmetric* if there is a time t_S such that $e(t+t_S)=e(t-t_S)$.

As it is quite clear, whereas time-reversal invariance is a property of dynamical equations (laws), time-symmetry is a property of a single solution (evolution) of a dynamical equation. Therefore, the time-reversal invariance of an equation does not imply the time-symmetry of its solutions: a time-reversal invariant law may be such that all or most of the possible evolutions relative to it are individually time-asymmetric. Huw Price [11] illustrates this point with the familiar analogy of a factory which produces equal numbers of left-handed and right-handed corkscrews: the production as a whole is completely unbiased, but each individual corkscrew is asymmetric.

In the context of general relativity, the concept of time-symmetry cannot be defined in so simple terms, but it can be formulated by means of the concept of time-isotropy:

Definition 12: A time-orientable spacetime (M,g) is *time-isotropic* if there is a diffeomorphism d of M onto itself which reverses the temporal orientations but preserves the metric g.

On the basis of this definition, the concept of time-symmetry for a spacetime which admits a global time can be characterized as follows:

Definition 13: A time-orientable spacetime (M,g) which admits a global time is *time-symmetric* with respect to some spacelike hypersurface Σ corresponding to a constant value of the global time, if it is time-isotropic and the diffeomorphism d leaves fixed the hypersurface Σ.

Intuitively this means that, from the hypersurface Σ, the spacetime *looks the same* in both temporal directions. Therefore, if a time-orientable spacetime endowed with a global time is time-asymmetric, we will not find a spacelike hypersurface Σ which splits the spacetime into two "halves", one the temporal mirror image of the other with respect to their intrinsic geometrical properties.

But, how does this time-asymmetry allow us to choose a time-orientation of the spacetime? As we have seen, in a time-orientable spacetime a continuous non-vanishing non-spacelike vector field $\gamma^{\mu}(x)$ can be defined all over the manifold. Up to this point, the universe is *time-orientable* but not yet *time-oriented*, because the distinction between $\gamma^{\mu}(x)$ and $-\gamma^{\mu}(x)$ is just conventional. Now time-asymmetry comes into play: in a time-asymmetric spacetime, *any* spacelike hypersurface Σ splits the manifold into two sections that are different to each other: the universe at one side of Σ is *substantially* different than the universe at the other side. We can choose any point $x_0 \in \Sigma$ and conventionally consider

4 The exception is the law that rules weak interactions. We will return on this point in the last section.

that the non-spacelike vector $\gamma^{\mu}(x_0)$ points towards one side of Σ and that $-\gamma^{\mu}(x_0)$ points towards the other side or vice versa: in any case we have established a substantial difference between $\gamma^{\mu}(x_0)$ and $-\gamma^{\mu}(x_0)$ since these vectors point to substantially different parts of the universe. We can conventionally call the direction of $\gamma^{\mu}(x_0)$ "future" and the direction of $-\gamma^{\mu}(x_0)$ "past" or vice versa, but in any case past is substantially different than future. Now we can extend this difference to the whole continuous fields $\gamma^{\mu}(x)$ and $-\gamma^{\mu}(x)$ obtained by making the global continuations of $\gamma^{\mu}(x_0)$ and $-\gamma^{\mu}(x_0)$ permitted by the definition of time-orientability: in this way, the time-orientation of the spacetime has been established. Since the field $\gamma^{\mu}(x)$ is defined all over the manifold, it can be used locally at each point x to define the future and the past semicones: for instance, if we have called the direction of $\gamma^{\mu}(x)$ "future", $C_+(x)$ contains $\gamma^{\mu}(x)$ and $C_-(x)$ contains $-\gamma^{\mu}(x)$. Summing up, in a time-asymmetric spacetime it is possible to distinguish between the two directions of time exclusively on de basis of geometrical considerations and with no reference to our particular perspective with regard to past and future.

The Generic Nature of the Arrow of Time

As Savitt [25] correctly points out, there are two different questions involved in the problem of the arrow of time:

- The "how possible" question: how is it possible to formulate a time-asymmetric model by means of time-reversal invariant laws?
- The "why probable" question: what reason is there to suppose that time-asymmetry is probable?

Up to this point it is quite clear that the "how possible" question has a straightforward answer: the time-reversal invariance of laws is not an obstacle to the construction of time-asymmetric models where the substantial distinction between the two directions of time can be established. However, the "why probable" question remains: it is necessary to supply an argument for the high probability of time-asymmetry. In this subsection we will address this second question by proving that, in time-orientable spacetimes endowed with a global time, *time-asymmetry is generic* to the extent that time-symmetric solutions of the universe equations have measure zero in the corresponding phase space.

We will begin with a simple case. Let us consider a time variable function $x(t)$ defined by a time-reversal invariant second order equation $\ddot{x}(t) = F[x(t)]$. By definition of time-reversal invariance, if $x_1 = x(t)$ is a solution of the equation, then $x_2 = x(-t)$ is also a solution. In the general case, the initial conditions at $t = 0$ of both solutions are $x_1(0) = x_2(0)$ and $\dot{x}_1(0) = -\dot{x}_2(0)$. But in the particular case in which $\dot{x}_1(0) = -\dot{x}_2(0) = 0$, both solutions $x_1(t)$ and $x_2(t)$ coincide, leading to a single solution $x_1(t) = x_2(-t) = x(t) = x(-t)$ which is time-symmetric with respect to the axis $t = 0$.

This means that whereas the space of the initial conditions corresponding to all the possible solutions is a two dimensional space defined by $(x(0), \dot{x}(0))$, the space of the initial conditions corresponding to time-symmetric solutions is a one dimensional space defined by $(x(0), \dot{x}(0) = 0)$ since in this case $\dot{x}(0) = -\dot{x}(0) = 0$. If the dynamical equation describes the time evolution of a typical system of classical mechanics, the time-symmetry conditions become $(q(0), p(0) = 0)$. Therefore, we can conclude either that the subset of time-symmetric solutions has measure zero in the set of all possible solutions (this claim cannot be asserted in spaces with no measure) or that the space of time-symmetric solutions is a proper subspace of the space of all possible solutions (this conclusion is valid also in infinite-dimensional spaces). Nevertheless, in both cases the central idea is the same: *time-asymmetry is generic*, time-symmetry is extraordinarily specific.

The argument developed for a simple case can be now extended to general relativity and spacetimes governed by the time-reversal invariant Einstein's field equations. Since time-orientability and existence of a global time are necessary conditions for defining the arrow of time, we will focus our attention in spacetimes describable by the ADM formalism, where the foliability of the spacetime is assumed[5]. In this context, the definition of time-symmetry for a time-orientable spacetime endowed with a global time function (Definition 13) implies that there exists a spacelike hypersurface Σ such that the extrinsic curvature vanishes at all of its points: $K_{ij} = 0$ $(i, j = 1,2,3$, $g = \det g_{ij})$ [26]. On the other hand, the initial conditions are given by the following Cauchy data: g_{ij} (that is, the geometry of Σ) and its time derivative $\dot{g}_{ij} = \dfrac{\partial g_{ij}}{\partial t}$ (where t is the global time function of the ADM metric). But these data amount to have g_{ij} and its conjugated momentum π^{ij}, which can be computed as (see [26] eq.(21.91)):

$$\pi^{ij} = g^{1/2}(g^{ij} g_{hk} K^{hk} - K^{ij}) \tag{6}$$

According to this equation, if the time-symmetry condition $K_{ij} = 0$ holds, then $\pi^{ij} = 0$. This means that the Cauchy data corresponding to spacetimes that are time-symmetric with respect to Σ are given by g_{ij} arbitrary and $\pi^{ij} = 0$ (in complete agreement with the simple case presented above where $q(0)$ is arbitrary and $p(0) = 0$). Let us denote by E the space of all the possible Cauchy data, $\{g_{ij}(x^1, x^2, x^3),\ \pi^{ij}(x^1, x^2, x^3)\}$, and by E_S the space of the Cauchy data that lead to time-symmetric geometries, $\{g_{ij}(x^1, x^2, x^3),\ \pi^{ij}(x^1, x^2, x^3) = 0\}$. Analogously to the previous case we can conclude that E_S is a proper

[5] This demonstration could be extended to more general spacetimes by using the framework of [23] (page 251), but this is not relevant to our case since we are interested in the problem of the arrow of time where the existence of a global time function is a necessary condition.

subspace of E and, therefore, *time-asymmetric spacetimes are generic* while time-symmetric spacetimes are extraordinarily specific[6].

As we can see, this global geometrical approach has not to face the problem of explaining overwhelmingly improbable initial conditions in order to account for the arrow of time. In contrast to the global entropic approach, from this perspective the arrow of time is a completely generic feature which does not require an exceptional fine-tuning of all the variables of the universe.

VI. Application to FLRW Models

In the previous section we have considered the general conditions necessary for the existence of a global and non-entropic arrow of time. In this section we will apply those conditions to the particular case of our universe.

As it is well known, on the basis of the cosmological principle, the (averaged) metric of the spacetime can be represented in the simple Robertson-Walker form:

$$ds^2 = dt^2 - a^2(t)\left(\frac{dr^2}{1-kr^2} + r^2 d\theta^2 + r^2 \sin^2\theta d\phi^2\right) \qquad (7)$$

With the Robertson-Walker metric, the Einstein's field equations can be solved: the corresponding solutions describe the isotropic and homogeneous FLRW models, which are the standard models of present day cosmology.

In the following subsections we will show that FLRW models satisfy the necessary conditions for the existence of a global and non-entropic arrow of time, and we will define such an arrow in this particular case.

Time-Orientability

FLRW models provide a simple answer to the question about the first condition for the existence of the arrow of time since they are time-orientable. Moreover, astronomical observations provide empirical evidence that makes implausible the non time-orientability of our spacetime. In particular, there is no astronomical observation of temporally inverted behavior in some (eventually very distant) region of the universe.

According to contemporary cosmology, at the decoupling time (400.000 years after the Big Bang) the universe was essentially composed of light elements like H and He (with traces of isotopes of these elements and of Li and Be) and was virtually free of ions. This matter condensed in stars, where heavy elements were formed. The explosion of these stars as supernovae scattered the heavy elements producing clouds which, in turn, condensed giving rise to a second generation of stars. Analogously, a third generation arose, and so on. We have an indirect evidence of the first generation (known as population III), appeared 200 millions

---

[6] Although we have discussed the case of empty universes, the demostration can be easily generalized to universes with a scalar field or an electromagnetic field [26], and to many other solutions of the Cauchy problem [27], [28].

years after the Big Bang, by the observed re-ionization of the universe which can be explained only by the presence of stars at that period [29]. Then, two new generations (populations II and I) followed. This is the history as manifest in all the universe.

- In this history, the evolutions of the different generations of supernovae always follow the same pattern (let's say, from birth to death), and there is no trace of a time-inverted pattern in the visible universe. This is a relevant fact in the context of our problem since the supernovae are the markers or standard candles used to measure the more distant galaxies ($z = 1$ and even more) [30].
- In turn, the peak in the quasars formation rate took place when the universe was 3.000 millions years old ($z \sim 2$), and this rate has always decreased since then (as shown by astronomical observations in all spatial directions). This fact can be considered as a good indication that the universe is time-orientable, at least up to the corresponding distances.
- Furthermore, the process of decoupling between matter and radiation occurred 400.000 years after the Big Bang ($z = 1000$, the limit of the visible universe in all spatial directions) and never happened again. This fact also counts as evidence for the time-orientability of the universe (since 400.000 years after the Big Bang up to the present).

Global Time and Cosmic Time

As we have seen, the existence of a global time is a precondition for the arrow of time. Nevertheless, cosmological models may have an even stronger temporal condition. In fact, a global time function does not yet permit to define the notion of simultaneity in an univocal manner and with physical meaning.

Let us consider the set of the worldline curves of all the particles of the universe, and a particular foliation defined by the global time function $t = \tau$, such that the worldline curves traverse all the hypersurfaces $\tau = const$. Now, let us endow the hypersurface corresponding to, say, $\tau = 0$ with a system of coordinates (x^i), with $i = 0$ to 3. Each worldline curve is labelled by the coordinates (x^1, x^2, x^3) of the point where it intersects the hypersurface $\tau = 0$. Now, we endow each hypersurface with $\tau = const$ with a system of coordinates (x^i) such that the point of intersection between the curve (x^1, x^2, x^3) and the hypersurface also has the coordinates (x^1, x^2, x^3), and this holds for all the worldline curves. In this way we have constructed a system of coordinates for the whole manifold which assigns the coordinates (τ, x^1, x^2, x^3) to each point in such a way that, given the curve (x^1, x^2, x^3), τ increases along it and all its points have the same value of the coordinates (x^i). In this coordinate system, the metric reads:

$$ds^2 = g_{00}\, d\tau^2 + 2g_{0i}\, d\tau\, dx^i + g_{ij}\, dx^i\, dx^j \tag{8}$$

where $g_{\mu\nu} = g_{\mu\nu}(\tau, x^1, x^2, x^3)$. It is quite clear that this procedure does not allow us to single out a preferred coordinate system for the whole manifold: we can construct as many coordinate systems as different foliations can be chosen; each foliation defines its corresponding simultaneity. In order to avoid ambiguities in the notion of simultaneity, we must choose a particular foliation. The foliation τ such that all the worldline curves are orthogonal to all the hypersurfaces $\tau = const$ is the proper choice, because orthogonality recovers the notion of simultaneity of special relativity for small regions (tangent hyperplanes) of the hypersurfaces $\tau = const$ (for the necessary conditions, see [26]). In this case, $g_{0i} = 0$ and, then, the metric reads:

$$ds^2 = g_{00}\, d\tau^2 + g_{ij}\, dx^i\, dx^j \tag{9}$$

Nevertheless, even if this condition selects a particular foliation, it permits that the proper time interval between two hypersurfaces of simultaneity be relative to the particular worldline considered for computing it. In fact, the proper time $\Delta\tau$ elapsed between τ_1 and τ_2 along the curve (x^1, x^2, x^3) can be computed as the distance Δs between the hypersurfaces $\tau = \tau_1$ and $\tau = \tau_2$ along such a curve:

$$\Delta\tau = \Delta s(\tau_1, \tau_2, x^1, x^2, x^3) = \int ds = \int [g_{00}(\tau, x^1, x^2, x^3)]^{1/2}\, d\tau \tag{10}$$

Since, in general, g_{00} is a function of (x^1, x^2, x^3), the proper time $\Delta\tau$ also depends on (x^1, x^2, x^3), that is, on the particular worldline curve used to compute it. If we want to avoid this situation, we must impose the constraint $g_{00} = g_{00}(\tau)$: when g_{00} is just a function of τ, the proper time interval between two hypersurfaces $\tau = \tau_1$ and $\tau = \tau_2$ is the same when measured on any orthogonal worldline curve. Therefore, it is possible to define a cosmic time t:

$$dt^2 = g_{00}(\tau)\, d\tau^2 \tag{11}$$

In this case, the metric results:

$$ds^2 = dt^2 - h_{ij}\, dx^i\, dx^j \tag{12}$$

Definition 14: When the metric of the spacetime (M, g) can be expressed as $ds^2 = dt^2 - h_{ij}\, dx^i\, dx^j$, t is the *cosmic time* and $h_{ij} = h_{ij}(t, x^1, x^2, x^3)$ is the three-dimensional metric of each hypersurface of simultaneity.

Only the existence of a cosmic time guarantees that all the processes of the universe can be coordinated by a single time that recovers the temporal structure of pre-relativistic physics, where the history of the universe as a whole can be univocally conceived as the temporal sequence of its instantaneous states.

These considerations can be easily applied to the case of FLRW models because eq.(7) is a particular case of eq.(12). Since FLRW models are spatially homogeneous and isotropic on the large scale, it is possible to find a family of spacelike hypersurfaces which can be labeled by the proper time of the worldlines that orthogonally thread through them: these labels define the cosmic time, which is represented by the variable t of the Robertson-Walker metric, and the scale factor a is a scalar only function of t. Therefore, to the extent that FLRW models have a cosmic time, *a fortiori* they satisfy the weaker condition of admitting a global time function.

Let us note that the existence of a cosmic time imposes a too strong condition to realistic and detailed cosmological models. For instance, a spacetime with cosmic time does not admit, for instance, a local Schwarzschild metric; this fact, in turn, precludes the occurrence of well known physical phenomena, like the existence of solar systems or the gravitational lensing effect. This means that FLRW models are valid only on a large scale and in an averaged sense, but must be modified when local cosmological phenomena have to be explained. Nevertheless, this kind of modifications does not count against the existence of a global and non-entropic arrow of time, since such an arrow requires that the spacetime admits a global time function and this condition holds even when that kind of local cosmological phenomena are taken into account.

It is interesting to emphasize how these considerations provide a further argument against the global entropic approach, which relies on the possibility of defining the entropy gradient of the universe, that is, the gradient of the entropy function $S(t)$. In the standard formulations, the variable t in the entropy function $S(t)$ denotes a time that plays the role of dynamical parameter in the evolution of the universe. This time should be the proper time measured by a clock fixed to the worldlines orthogonal to the foliation. But $S(t)$ would result not well defined if such a time t changed from one trajectory to the other. As a consequence, only the existence of a cosmic time guarantees that the temporal structure of pre-relativistic physics can be recovered and the entropy function $S(t)$ can be satisfactorily defined. This means that, in order to define the arrow of time, the global entropic approach imposes a condition so strong that precludes well known cosmological phenomena.

Time-Asymmetry

When, as in the case of FLRW models, the spacetime admits a cosmic time, the issues of time-reversal invariance and time-asymmetry can be presented in more familiar terms than those used in the previous sections. In fact, when the spacetime has a cosmic time, Einstein's field equations are time-reversal invariant in the sense that if the $h_{ij}(t,x^1,x^2,x^3)$ is a solution, $h_{ij}(-t,x^1,x^2,x^3)$ is also a solution. But the time-reversal invariance of these equations does not prevent us from describing a time-asymmetric universe whose spacetime is asymmetric regarding its geometrical properties along the cosmic time. In this case, the concept of time-symmetry can be defined as follows:

Definition 15: A time-orientable spacetime (M,g) which admits a cosmic time t is *time-symmetric* with respect to some spacelike hypersurface $t = t_S$, where t_S is a constant, if it is time-isotropic and the diffeomorphism d leaves fixed the hypersurface $t = t_S$.

On the basis of these considerations, it is easy to see that in FLRW models the time-symmetry of spacetime may manifest itself in two different ways according to whether the universe has singular points in one or in both temporal extremities[7]. Big Bang-Big Chill universes are manifestly time-asymmetric: since the scale factor $a(t)$ increases with the cosmic time t, there is no hypersurface $t = t_S$ from which the spacetime looks the same in both temporal directions. In Big Bang-Big Crunch universes, on the contrary, $a(t)$ has a maximum value: therefore, the spacetime might be time-symmetric about the time of maximum expansion. These are the cases of some FLRW models with dust and radiation.

In those models (*e.g.* in the particular case of inflationary models) different fields can represent the matter-energy of the universe. Many interesting results have been obtained, for instance, by modeling matter-energy as a set of scalar fields $\phi_k(t)$: homogeneity is retained and the time-reversal invariance of the field equations is given by the fact that, if $[a(t), \phi_k(t)]$ is a solution, $[a(-t), \phi_k(-t)]$ is also a solution. In these cases, if there is a time t_{ME} of maximum expansion, the scale factor $a(t)$ may be such that $a(t_{ME}+t) \neq a(t_{ME}-t)$ (see, for instance, the models in [31]); as we will see in the next subsection, this is the generic situation. This means that a Big Bang-Big Crunch universe may be a time-asymmetric object with respect to the metric of the spacetime, and this asymmetry, essentially grounded on geometrical considerations, allows us to distinguish between the two directions of the cosmic time, independently of entropic considerations.

The Generic Nature of the Arrow of Time

Although we have proved the generic nature of the arrow of time in the general case of a time-orientable spacetime endowed with a global time function, it is interesting to see how the argument becomes a very simple demonstration in the case of FLRW models. Of course, the argument is superfluous in the case of Big Bang-Big Chill universes, where the temporal asymmetry is manifest. Then, let us consider a Big Bang-Big Crunch FLRW universe with radius a and matter represented by a neutral scalar field ϕ. The dynamical variables of the model are $a, \dot{a}, \phi, \dot{\phi}$, which satisfy a generic Hamiltonian constraint:

$$H(a, \dot{a}, \phi, \dot{\phi}) = 0 \tag{13}$$

that reduces the dimension of phase space from 4 to 3, since we can express:

$$a = f(\dot{a}, \phi, \dot{\phi}) \tag{14}$$

[7] This depends on the values of the factor k and of the cosmological constant Λ.

Therefore, the state of this universe can be represented by a point in a three dimensional phase space with coordinates $\dot{a}, \phi, \dot{\phi}$. If we want to obtain a time-symmetric continuous[8] solution such that $a \geq 0$[9], there must be a time t_S regarding to which a is symmetric:

$$a(t_S + t) = a(t_S - t) \quad \text{and} \quad \dot{a}(t_S) = 0 \tag{15}$$

In order to obtain complete time-symmetry, ϕ must also be symmetric about t_S. There are two cases: even symmetry,

$$\phi(t_S + t) = \phi(t_S - t) \quad \textit{and} \quad \dot{\phi}(t_S) = 0 \tag{16}$$

and odd symmetry,

$$\phi(t_S + t) = -\phi(t_S - t) \quad \textit{and} \quad \phi(t_S) = 0 \tag{17}$$

This means that time-symmetric trajectories necessarily pass trough the axes $(0, \phi, 0)$ or $(0, 0, \dot{\phi})$ of the phase space. From these "initial" conditions we can propagate, using the evolution equations, the corresponding trajectories; this operation will produce two surfaces that contain the trajectories with at least one point of symmetry, that is, two surfaces containing all the possible time-symmetric trajectories. Since both surfaces have dimension 2 in a 3-dimensional phase space, their usual Liouville measure is zero, and also any measure absolutely continuous with respect to it. If we conceive the Liouville measure on the phase space as a measure of probability, this argument shows that time-symmetric universes have probability zero with respect to the reference class of all possible universes compatible with a Big Bang-Big Crunch FLRW model with a scalar field.

The argument be easily generalized to the case where ϕ has many components, or to the case of many fields with many components. Some of these fields may be fluctuations of the metric: in this case, we must Fourier transform the equations, and this would allow us to reproduce the theorem only with t functions. Since time-reversal invariance also holds in the classical statistical case, the theorem can be also demonstrated in this case. And also in the quantum case, albeit some quantum gravity problems like time definition [32].

---

[8] We will disregard non-continuous solutions since normally information do not pass through discontinuities and we are only considering connected universes where information can go from a point to any other timelike connected point.

[9] As only a^2 appears in a FRW metric, we will consider just the case *a 0* since the point *a=0* is actually a singularity that cuts the time evolution.

VII. From Geometry to Energy Flow

The Global Arrow of Time as Energy Flow

As we have seen, in a time-orientable spacetime having a global time, time-asymmetry establishes a substantial difference between the two temporal directions represented by the continuous non-vanishing non-spacelike vector fields $\gamma^{\mu}(x)$ and $-\gamma^{\mu}(x)$. Then, we can define a substantially grounded time-orientation by deciding that, for instance, $\gamma^{\mu}(x)$ points to the future; in this case, any vector of the vector field $\gamma^{\mu}(x)$ represents the arrow of time. However, up to this point $\gamma^{\mu}(x)$ was characterized in mathematical terms as the vector field that must exist for the time-orientability of spacetime. The question now is whether the arrow of time can be defined in a physical way, that is, by means of a mathematical object that can be interpreted not only geometrically but also in terms of the more familiar magnitudes of physics.

As it is well known, the energy-momentum (or gravitational stress) tensor $T_{\mu\nu}$ represents the density and the flow of energy and momentum at each point of the spacetime. Then, it would be desirable to define the vector field $\gamma^{\mu}(x)$ in terms of $T_{\mu\nu}$ in order to endow it with a physical meaning[10]. Although this task cannot be accomplished in a completely general case, it is possible to define the arrow of time in terms of $T_{\mu\nu}$ if two conditions are satisfied: (i) the density of matter-energy is non-zero everywhere, and (ii) the energy-momentum tensor satisfies the *dominant energy condition* everywhere. The first condition means that there is perfect vacuum in no region of the universe; this seems to be a physically realistic situation since in any region of the universe there exists, at least, blackbody cosmic radiation. The second condition is based on the following definition (see [23], [33]):

Definition 16: The energy-momentum tensor satisfies the *dominant energy condition* if, in any orthonormal basis, the energy component dominates the other components of $T_{\alpha\beta}$:

$$T^{00} \geq \left|T^{\alpha\beta}\right| \qquad \textit{for each } \alpha, \beta$$

where the first letters of the Greek alphabet are reserved for the indices of the orthonormal bases.

This means that, to any local observer, the matter-energy density T^{00} appears non-negative and the energy flow $T^{0\mu}$ is non-spacelike. The dominant energy condition does not impose a very strong constraint, since it holds for almost all known forms of ordinary matter. At present, a non-negative pressure cosmological term of $T_{\mu\nu}$ seems to exist in the universe. Even though this term does not satisfy the dominant energy condition, it can be considered as a background term, irrelevant for the local issues we are considering here. To the extent that the ordinary matter term does satisfy the dominant energy condition, we can include the non-

---

[10] Of course, this can be done only in the regions of the spacetime where $T_{\mu\nu} \neq 0$. In the regions where $T_{\mu\nu}=0$, we are forced to use the continuous non-vanishing non-spacelike vector field γ^{μ}, as explained in Section V.

negative pressure term in the l.h.s. of the Einstein's equations as $\Lambda g_{\mu\nu}$, and the resulting r.h.s. will satisfy the dominant energy condition..

Let us consider a continuous orthonormal basis field $\{V^{\mu}_{(\alpha)}(x)\}$ (a tetrad or *vierbein*) consisting of four unitary vectors $(V^{\mu}_{(0)}(x), V^{\mu}_{(1)}(x), V^{\mu}_{(2)}(x), V^{\mu}_{(3)}(x))$. In this basis, $g_{\mu\nu}V^{\mu}_{(\alpha)}V^{\nu}_{(\beta)} = \eta_{\alpha\beta}$ are the coordinates of the local Minkowski metric tensor, and $T_{\mu\nu}V^{\mu}_{(\alpha)}V^{\nu}_{(\beta)} = T_{\alpha\beta}$ are the coordinates of the energy-momentum tensor. Then, $T^{0\alpha}V^{\mu}_{(\alpha)}$ can be conceived as a vector representing the energy flow, whose coordinates in the basis $\{V^{\mu}_{(\alpha)}(x)\}$ are the $T^{0\alpha}$. Now, if $T^{00} \geq \left|T^{\alpha\beta}\right|$, then $T^{00} \geq \left|T^{0\alpha}\right|$. In turn, $T^{00} \geq \left|T^{0\alpha}\right|$ implies that $T^{0\alpha}V^{\mu}_{(\alpha)}$ is non-spacelike. On the other hand, if the manifold and the orthogonal basis field are continuous, $g_{\mu\nu}$ is continuously defined over it and, provided that the derivatives of $g_{\mu\nu}$ are also continuous, $T^{\mu\nu}$ ($T^{\alpha\beta}$) and, then, $T^{0\mu}$ ($T^{0\beta}$) are also continuously defined all over the manifold. Therefore, it seems that, in the regions where the density of matter-energy is non-zero, we have found **a** physical correlate of the continuous non-vanishing non-spacelike vector field $\gamma^{\mu}(x) = T^{0\alpha}(x)V^{\mu}_{(\alpha)}(x)$. The trouble with this conclusion is that $T^{0\alpha}V^{\mu}_{(\alpha)}$ is not strictly a vector, since it is not transformed as a vector by the Lorentz transformations. Strictly speaking, at each point x of the spacetime $T^{0\alpha}(x)V^{\mu}_{(\alpha)}(x)$ is a tetra-magnitude which represents the energy flow only in the basis $\{V^{\mu}_{(\alpha)}(x)\}$; thus, it seems that it cannot directly play the role of $\gamma^{\mu}(x)$ as initially desired.

Nevertheless, the fact that the energy flow cannot be represented by a vector is not an obstacle to define the arrow of time in terms of such a flow. The dominant energy condition poses a *covariant* condition: if the energy flow is non-spacelike in a reference frame, it is non-spacelike in all reference frames. This means that, no matter which orthonormal basis $\{V^{\mu}_{(\alpha)}(x)\}$ is chosen, the energy flow in that basis, represented by $T^{0\alpha}(x)V^{\mu}_{(\alpha)}(x)$, can be used to define the arrow of time. In fact, we can conventionally decide that, at each point x of the spacetime, the future light semicone $C_{+}(x)$ is the one that contains the energy flow whatever basis was chosen: in this way, the time-orientation is defined for the whole spacetime.

Of course, the fact that the time-orientation of a spacetime can be defined by means of the energy flow does no imply that time-asymmetry becomes superfluous. Given a spacetime with its own geometry, something analogous to the case of the pair $\gamma^{\mu}(x)$ and $-\gamma^{\mu}(x)$ happens with the energy flow: in any orthonormal basis $\{V^{\mu}_{(\alpha)}(x)\}$, for each positive $T^{0\alpha}(x)V^{\mu}_{(\alpha)}$ there is a negative $-T^{0\alpha}(x)V^{\mu}_{(\alpha)}$ that can also adequately represent the energy flow, since the difference between them is, *a priori*, just conventional. It is the time-asymmetry of the spacetime what establishes the substantial difference between both pseudovectors. Once again, we can conventionally call the temporal direction of $-T^{0\alpha}(x)V^{\mu}_{(\alpha)}$ "past" and the temporal direction of $T^{0\alpha}(x)V^{\mu}_{(\alpha)}$ "future" or vice versa, but in any case past is substantially different than future. The usual convention in physics consists in calling the temporal direction of the positive energy flow "future". This means that, no matter which orthonormal basis we use, at any point x of the spacetime $T^{0\alpha}(x)V^{\mu}_{(\alpha)}(x)$ belongs to the future light semicone $C_{+}(x)$: the energy temporally flows towards the future for any observer. But the relevant point is that this sentence acquires a non-conventional meaning only when the *substantial difference* between past and future has been previously established by the time-asymmetry of the spacetime.

From the Global Arrow to the Local Arrow

As we have seen, according to the usual terminological convention, the future light semicone $C_{+}(x)$ at each point x of the spacetime is defined by the positive energy flow $T^{0\alpha}(x)V^{\mu}_{(\alpha)}(x)$ at this point. But, does $T^{0\alpha}$ really represent the energy flow as conceived by local physics? Let us remember that $T_{\mu\nu}$ satisfies the "conservation" equation:

$$\nabla_{\mu} T^{\mu\nu} = 0 \tag{18}$$

However, this is not a true conservation equation since ∇_{μ} is a covariant derivative. The usual conservation equation with ordinary derivative reads:

$$\partial_{\mu} \tau^{\mu\nu} = 0 \tag{19}$$

where $\tau_{\mu\nu}$, the energy-momentum pseudo-tensor (since it is not a proper tensor), is defined as [26]:

$$\tau^{\mu\nu} = T^{\mu\nu}_{eff} = T^{\mu\nu} + t^{\mu\nu} \tag{20}$$

The correction term $t_{\mu\nu}$ reads:

$$t_{\mu\nu} = \frac{1}{16\pi}\left[L g_{\mu\nu} - \frac{\partial L}{\partial g^{\mu}_{\kappa\lambda}} g_{\kappa\lambda,\nu} \right] \tag{21}$$

where L is the system's gravitational Lagrangian[11]. The correction factor $t_{\mu\nu}$ is an homogeneous and quadratic function of the connection $\Gamma^{\lambda}_{\nu\mu}$ [34], precisely:

$$t^{\mu\nu} = \frac{1}{16\pi k}\Big[\left(2\Gamma^{\chi}_{\lambda\rho}\Gamma^{\sigma}_{\chi\sigma} - \Gamma^{\chi}_{\lambda\sigma}\Gamma^{\sigma}_{\rho\chi} - \Gamma^{\chi}_{\lambda\chi}\Gamma^{\sigma}_{\rho\sigma}\right)(g^{\mu\lambda}g^{\nu\rho} - g^{\mu\nu}g^{\lambda\rho}) + \\ + g^{\mu\lambda}g^{\rho\chi}\left(\Gamma^{\nu}_{\lambda\sigma}\Gamma^{\sigma}_{\rho\chi} + \Gamma^{\nu}_{\rho\chi}\Gamma^{\sigma}_{\lambda\sigma} - \Gamma^{\nu}_{\chi\sigma}\Gamma^{\sigma}_{\lambda\rho} - \Gamma^{\nu}_{\lambda\rho}\Gamma^{\sigma}_{\chi\sigma}\right) + \\ + g^{\nu\lambda}g^{\rho\chi}\left(\Gamma^{\mu}_{\lambda\sigma}\Gamma^{\sigma}_{\rho\chi} + \Gamma^{\mu}_{\rho\chi}\Gamma^{\sigma}_{\lambda\sigma} - \Gamma^{\mu}_{\chi\sigma}\Gamma^{\sigma}_{\lambda\rho} - \Gamma^{\mu}_{\lambda\rho}\Gamma^{\sigma}_{\chi\sigma}\right) + \\ + g^{\lambda\rho}g^{\chi\sigma}\left(\Gamma^{\mu}_{\lambda\chi}\Gamma^{\nu}_{\rho\sigma} - \Gamma^{\mu}_{\lambda\rho}\Gamma^{\nu}_{\chi\sigma}\right)\Big] \tag{22}$$

Now we can consider the coordinates $\tau^{0\mu}$, which satisfy a usual conservation equation:

$$\partial_{\mu}\tau^{0\mu} = \partial_0\tau^{00} + \partial_i\tau^{0i} = 0 \tag{23}$$

where τ^{00} represents the energy density and the τ^{0i} are the three components of the spatial energy flow (the Poynting vector).

Analogously to the case of the energy-momentum tensor, in any orthonormal basis $\left\{V^{\mu}_{(\alpha)}\right\}$, $\tau^{0\alpha}V^{\mu}_{(\alpha)}$ is not a vector since it is not transformed as a vector by the Lorentz transformations. But in contrast to that case, the dominant energy condition cannot be expressed in terms of $\tau^{\mu\nu}$; this means that we cannot guarantee that $\tau^{0\alpha}V^{\mu}_{(\alpha)}$ is non-spacelike in any orthonormal basis. However, this can be done in the particular case of *local inertial frames*. In fact, in a local inertial frame, $\Gamma^{\lambda}_{\nu\mu} = 0$; thus, $t_{\mu\nu} = 0$ (see eq.(22)) and $\tau^{\mu\nu}$ = $T^{\mu\nu}$ (see eq.(20)). Therefore, in this case the dominant energy condition implies that:

$$\tau^{00} \geq \left|\tau^{\alpha\beta}\right| \qquad \textit{for each } \alpha,\beta \tag{24}$$

As a consequence, in the basis $\left\{W^{\mu}_{(\alpha)}\right\} = \left\{W^{\mu}_{(0)}, W^{\mu}_{(1)}, W^{\mu}_{(2)}, W^{\mu}_{(3)}\right\}$ of the local inertial frame, the energy flow, represented by $\tau^{0\alpha}W^{\mu}_{(\alpha)}$ and satisfying the usual conservation equation, is non-spacelike.

Although this result cannot be generalized to all the reference frames of the whole spacetime, it is relevant in local contexts since, in small regions of the spacetime, the metric

[11] Since both sides of the Einstein's equations satisfy the covariant conservation law (18), the same correction term $t_{\mu\nu}$ can be used in both sides of such equations in order to obtain the corresponding usual conservation law (19). Therefore, such a term can be computed by means of the gravitational Lagragian that defines the l.h.s of those equations. This means that the correction term $t_{\mu\nu}$, given by eq.(21), can be adequately defined either when matter is present or in the vacuum.

tends to the Minkowski form in local inertial frames[12]. In fact, any local region of the spacetime can be approximated by the tangent Minkowski space (with orthonormal basis $\left\{V_{(\alpha)}^{\mu}\right\}$) at any point of that region. If $\left\{W_{(\alpha)}^{\mu}\right\}$ is the basis of an inertial frame on this flat tangent space, $\tau^{0\alpha}W_{(\alpha)}^{\mu}$ represents the non-spacelike *local energy flow* satisfying the usual conservation equation[13]. Moreover, at each point x of the local region, the local energy flow $\tau^{0\alpha}(x)W_{(\alpha)}^{\mu}(x)$ belongs to the same light semicone as the one to which the global energy flow $T^{0\alpha}(x)V_{(\alpha)}^{\mu}(x)$ belongs. Therefore, if we have adopted the usual physical convention in the global level, we can also meaningfully say that future is the temporal direction of the positive local energy flow: the local flow emitted at x points to the future, that is, belongs to the light semicone $C_+(x)$.

This result is particularly relevant because local inertial frames are the reference frames in which the non-relativistic local theories of physics are valid. In turn, ordinary quantum field theory in flat spacetime, although relativistic, must be considered as locally formulated in a local inertial frame. This means that the local energy flow directed towards the future is the flow of energy as conceived by this kind of theories, where energy satisfies the usual conservation law expressed by means of ordinary derivatives. Summing up, $\tau^{0\mu}$ inherits the time-orientation defined at the global level and, to the extent that it has a local physical meaning, it not only transfers the global arrow of time to local contexts, but also translates the global arrow into a usual magnitude of local physical theories.

Arrow of Time and Symmetry Transformations

As we have seen, the vector $\tau^{0\mu}$ introduced above *is always non-spacelike* and, therefore, its direction can represent the arrow of time when the spacetime is time-asymmetric. However, we know that, given the time-reversal invariance of Einstein's field equations, if we obtain a solution where $\tau^{0\mu}$ can be defined, we will also obtain another solution where $-\tau^{0\mu}$ can be defined. In other words, we can always obtain two solutions of Einstein's field equations: each one of them is the temporal mirror image of the other and both are possible with respect to those equations. At this point, the ghost of symmetry threatens again: it seems that we are committed to supplying a non-conventional criterion for picking out one of both nomologically admissible solutions, one the temporal mirror image of the other. Nevertheless, as we will see, the threat is not so serious as it seems.

As we have seen, in the case of a time-reversal invariant equation, if $e(t)$ is a solution, $\mathbf{T}e(t)$ is also a solution. The point is to understand the meaning of the time-reversal transformation $\mathbf{T}$. Under the active interpretation, a symmetry transformation corresponds to

[12] Near each point x_0, the metric can be approximated with the metric of the local free inertial frame as: $ds^2 = (1 - R_{0l0m}x^l x^m)dt^2 - \left(\frac{4}{3}R_{0ljm}x^l x^m\right)dtdx^j + \left(\delta_{ij} - \frac{1}{3}R_{iljm}x^l x^m\right)dx^i dx^j + 0(|x^j|^3)dx^\alpha d^\beta$ (see [26] eq.(13.73)). Therefore, locally in the inertial frame at x_0, we can guarantee that $\tau^{00} \geq |\tau^{0i}|$.

a change from one system to another; under the passive interpretation, a symmetry transformation consists in a change of the point of view from which the system is described. The traditional position about symmetries assumes that, in the case of discrete transformations as time-reversal or spatial reflection, only the active interpretation makes sense: an idealized observer can rotate himself in space in correspondence with the given spatial rotation, but it is impossible to "rotate in time" (see [19], [35]). Of course, this is true when the idealized observer is immersed in the same spacetime as the observed system. But when the system is the universe as a whole, we cannot change our spatial perspective with respect to the universe: it is as impossible to rotate in space as to rotate in time. However, this does not mean that the active interpretation is the correct one: the idea of two identical universes, one translated in space or in time regarding the other, has no meaning. This shows that both interpretations, when applied to the universe as a whole, collapse into conceptual nonsense.

In fact, in cosmology symmetry transformations are neither given an active nor a passive interpretation.

Definition 17: Two mathematical models for the universe, defined by (M,g) and (M',g'), are taken to be equivalent if they are *isometric*, that is, if there is a diffeomorphism $\theta : M \to M'$ which carries the metric g into the metric g' (see [23]).

Since symmetry transformations are isometries, two models related by a symmetry transformation (in particular, time-reversal) are considered equivalent descriptions of one of the same universe. Therefore, it is not necessary to find a non-conventional criterion for selecting one of the two nomologically admissible solutions related by time-reversal to the extent that both are descriptions of a single possible universe.

These considerations point to the absolute character of the arrow of time embodied in $\tau^{0\mu}$. This vector (in particular, its direction) is rigidly linked to the spacetime on which it is defined. To change $\tau^{0\mu}$ by $-\tau^{0\mu}$ amounts to change the model by its temporal mirror image; but, as we have shown, this move has no physical meaning since both models are merely conventionally different descriptions of the same universe.

Let us note that we have not used the term "future" in the present argument. If we adopt the usual terminology, we will call the time direction of the energy flow "future": in this case, we can say that the vector $\tau^{0\mu}$ points to the future. Nevertheless, it is worth remembering that "past" and "future" are words coming from our everyday language but they do not belong to physical theories. Then, the choice of saying that $\tau^{0\mu}$ points to future is conventional: we can replace "future" by "past" and nothing changes. What remains is the absolute and substantial nature of the arrow of time defined by the unchangeable direction of $\tau^{0\mu}$.

[13] In fact, $\partial_\mu(\tau^{0\alpha}W^{\mu}_{(\alpha)}) = \partial_\mu \tau^{0\mu} = 0$.

VIII. Time Symmetric Twins: Breaking the Symmetry

As we have seen in Section III, the Ehrenfests objected Gibbs' approach by pointing out that, if entropy is defined in statistical terms on the basis of the underlying time-reversal invariant classical mechanics, the increase of the entropy towards the future is always matched by a similar one in the past of the system. It is interesting to note that this old discussion can be generalized to the case of any kind of evolution arising from local time-reversal invariant laws. In fact, if $e(t)$ is a solution of a time-reversal invariant equation, $\mathbf{T}e(t)$ is also a solution. In general, any time-reversal invariant law gives rise to what we will call "*time-symmetric twins*" (see [2]), that is, two mathematical structures symmetrically related by the time-reversal transformation $\mathbf{T}$: each "twin", which in some cases represents an irreversible evolution, is the temporal mirror image of the other "twin". The traditional example of time-symmetric twins is given by electromagnetism, where dynamical equations always have advanced and retarded solutions, respectively related with incoming and outgoing states in scattering situations, whose mathematical structure is rigorously defined as described by Lax-Phillips scattering theory [36]. Moreover, in Lax-Phillips theory outgoing states (retarded solutions) correspond to Hardy classes from below (anticipating the results of the next subsection).

From the viewpoint of the local theory that produces the time-symmetric twins, the difference between them is only conventional: both twins are nomologically possible with respect to the theory. The traditional arguments for discarding one of the twins and retaining the other invoke time-asymmetric notions which are not justified in the context of the local theory. For instance, the retarded nature of radiation is usually explained by means of *de facto* arguments referred to initial conditions: advanced solutions of wave equations correspond to converging waves that require a "miraculous" cooperative emitting behavior of distant regions of space at the temporal origin of the process.

It seems quite clear that this kind of arguments, even if admissible in the discussions about irreversibility, are not legitimate in the context of the problem of the arrow of time to the extent that they put the arrow "by hand" by presupposing the difference between the two directions of time from the beginning. In other words, they violate the "nowhen" requirement of adopting an atemporal perspective purged of temporal intuitions like those related with the asymmetry between past and future or between initial and final conditions. Therefore, from an atemporal standpoint, the challenge consists in supplying a non-conventional criterion, based on theoretical arguments, for choosing one of the time-symmetric twins as the physically meaningful one: such a criterion will establish a substantial difference between the two members of the pair.

The desired criterion, that can be legitimately supplied neither by the local theory nor by our pretheoretical intuitions, is given by the local energy flow expressed in terms of $\tau^{0\mu}$. In fact, once we have conventionally decided to call the direction of the positive energy flow "future", at each spatio-temporal point x of a small region of the spacetime the local energy flow defines the future light semicone: the energy emitted at x must be contained in $C_+(x)$. Therefore, in any case the twin corresponding to this kind of energy flow is the member of the pair that must be retained as physically meaningful to represent the future directed twin. In the case of electromagnetism, only retarded solutions fulfill the required condition, since they describe outgoing (spontaneously evolving) states which are obtained by means of energy coming from the past and provide energy toward the future. In the following subsections, we will analyze in detail the breaking of symmetry in two fundamental theories of physics: quantum mechanics and quantum field theory.

Irreversible Quantum Mechanics

The so-called irreversible quantum mechanics is based on the use of rigged Hilbert spaces, due to the ability of this formalism to model irreversible physical phenomena such as decay or scattering processes ([3], [37]). The general strategy consists in introducing two subspaces, Φ_- and Φ_+, of the Hilbert space H. The vectors $|\phi\rangle$ of the subspaces Φ_- and Φ_+ are characterized by the fact that their projections $\langle\omega|\phi\rangle$ on the eigenstates of energy are functions of the Hardy class from above and from below respectively. These subspaces yield two rigged Hilbert spaces:

$$\Phi_- \subset H \subset \Phi_-^\times \qquad\qquad \Phi_+ \subset H \subset \Phi_+^\times \tag{25}$$

where $\Phi_-^\times$ and $\Phi_+^\times$ are the anti-dual of the spaces Φ_- and Φ_+ respectively.

It is quite clear that, up to this point, this general strategy amounts to obtain two time-symmetric twins from a time-reversal invariant theory. In fact, quantum mechanics formulated on a Hilbert space H is time-reversal invariant since:

$$KH = H \tag{26}$$

where K is the Wigner antilinear time-reversal operator. But if quantum mechanics is formulated on spaces $\Phi_\pm$, it turns out to be a non time-reversal invariant theory since:

$$K\Phi_\pm = \Phi_\mp \tag{27}$$

Moreover, in the analytical continuation of the energy spectrum of the system's Hamiltonian, there exists at least a pair of complex conjugate poles, one in the lower half-plane and the other in the upper half-plane of the complex plane. Such poles correspond to a pair of Gamov vectors:

$$|f_0^-\rangle \in \Phi_-^\times \qquad\qquad |f_0^+\rangle \in \Phi_+^\times \tag{28}$$

These vectors are proposed to describe irreversible processes and are taken to be the representation of the exponentially growing and decaying part of resonant unstable states, respectively ([4], [38]). But the symmetric position of the poles with respect to the real axis in the complex plane is a clear indication of the fact that Gamov vectors are also a case of time-symmetric twins.

In his detailed description of scattering processes, Arno Bohm breaks the symmetry between the twins by appealing to the so-called "*preparation-registration arrow of time*", expressed by the slogan "No registration before preparation" [39]. The key idea behind this proposal is that observable properties of a state cannot be measured until the state acting as a bearer of these properties has been prepared. For instance, in a scattering process, it makes no sense to measure the scattering angle until a state is prepared by an accelerator. On this basis, Bohm proposes the following *interpretational postulate*: the vectors $|\varphi\rangle \in \Phi_-$ represent the states of the system and the vectors $|\psi\rangle \in \Phi_+$ represent the observables of the system in the sense that observables are obtained as $O = |\psi\rangle\langle\psi|$. The time $t = 0$ is considered as the time at which preparation ends and detection begins. The preparation-registration arrow imposes the requirement that the energy distribution produced by the accelerator, represented by $\langle\omega|\psi\rangle$, be zero for $t > 0$, and the energy distribution of the detected state, represented by $\langle\omega|\psi\rangle$, be zero for $t < 0$:

$$\begin{array}{ll} |\varphi\rangle \in \Phi_- & \langle\omega|\psi\rangle = 0 \quad \text{for } t > 0 \\ |\psi\rangle \in \Phi_+ & \langle\omega|\psi\rangle = 0 \quad \text{for } t < 0 \end{array} \tag{29}$$

The time evolution, traditionally represented by the group U_t on the Hilbert space H, is here represented by two semigroups U_t^- and U_t^+: (i) U_t^- is U_t restricted to Φ_- and, then, it is valid only for $t > 0$, and (ii) U_t^+ is U_t restricted to Φ_+ and, then, it is valid only for $t < 0$ [40].

$$\begin{array}{ll} U_t^- : \Phi_- \to \Phi_- & |\varphi(t)\rangle = U_t^- |\varphi_0\rangle \quad \text{for } t > 0 \\ U_t^+ : \Phi_+ \to \Phi_+ & |\psi(t)\rangle = U_t^+ |\psi_0\rangle \quad \text{for } t < 0 \end{array} \tag{30}$$

As a consequence, the two Gamov vectors $|f_0^-\rangle \in \Phi_-^\times$ and $|f_0^+\rangle \in \Phi_+^\times$ turn out to be representations of growing and decaying states respectively: the evolution of the *growing Gamov vector* $|f_0^-\rangle$ can be defined only for $t < 0$, and the evolution of the *decaying Gamov vector* $|f_0^+\rangle$ can be defined only for $t > 0$.

As we can see, Bohm's approach breaks the symmetry between the time-symmetric twins by means of an interpretational postulate based on the preparation-registration arrow. Of course, this strategy supplies a solution to the problem of irreversibility to the extent that it permits the representation of irreversible growing and decaying processes. However, it does not offer a theoretical way out of the problem of the arrow of time, since the preparation-registration arrow is introduced as a postulate of the theory. In fact, such an arrow presupposes the distinction between past and future from the very beginning: in the past ($t < 0$) the system is prepared, in the future ($t > 0$) the system is measured, and $|f_0^-\rangle$ and $|f_0^+\rangle$ represent the corresponding growing and decaying processes respectively, both evolving toward the future. Furthermore, the inversion of the interpretational postulate would restore the time-symmetry: we could consider that Φ_+ is the space of states and Φ_- is the space of vectors by means of which the observables are obtained; in this case we would obtain the temporal mirror image of the original theory, where $|f_0^+\rangle$ and $|f_0^-\rangle$ represent growing and decaying states respectively, both evolving toward the past. In other words, the two possible interpretational postulates lead to two non time-reversal invariant theories, one the temporal mirror image of the other: they are also time-symmetric twins. Bohm's strategy of choosing the future directed version of the postulate introduces the arrow of time from the very beginning only on the basis of pretheoretical considerations.

Nevertheless, it is possible to find a theoretical justification for the preparation-registration arrow and, as a consequence, for choosing the future directed version of the interpretational postulate: the breaking of the symmetry between the time-symmetric twins is supplied by the local energy flow represented by $\tau^{0\mu}$. The preparation of the states $|\varphi\rangle \in \Phi_-$ acting as bearers of properties requires energy coming from other processes. Since the energy flow comes from the past and goes to the future, the states are prepared by means of energy coming from the past, that is, from previous processes; the growing Gamov vector $|f_0^-\rangle$ represents precisely the growing process occurring at $t<0$, which absorbs the energy coming from the past. On the other hand, the registration of the observable properties represented by $|\psi\rangle \in \Phi_+$ provides energy to other processes. Again, since the energy flow comes from the past and goes to the future, the measurement of observables emits energy toward the future, that is, provides energy to latter processes; the decaying Gamov vector $|f_0^+\rangle$ represents precisely the decaying process occurring at $t>0$, which emits energy toward the future. Summing up, the local energy flow coming from the past and directed to the future provides the criterion for selecting the future directed version of Bohm's interpretational postulate, and turns quantum mechanics into a non time-reversal invariant theory without the introduction of pretheoretical assumptions.

The Non Time-Reversal Invariance of Quantum Field Theory

In any of its versions, axiomatic quantum field theory includes a non time-reversal invariant postulate (see [41], [42], [43]), which states that the spectrum of the energy-momentum operator P^μ is confined to a future light semicone, that is, its eigenvalues p^μ satisfy:

$$p^2 \geq 0 \qquad\qquad p^0 \geq 0 \tag{31}$$

This postulate says that, when we measure the observable P^μ, we obtain a *non-spacelike classical* p^μ *contained in a future semicone*, that is, a semicone belonging to C_+.

It is clear that condition $p^0 \geq 0$ selects one of the elements of the time-symmetric twins $p^0 \geq 0$ and $p^0 \leq 0$ that would arise from the theory in the absence of the non time-reversal invariant postulate: by means of this postulate, quantum field theory becomes a non time-reversal invariant theory. In turn, since quantum field theory, being both quantum and relativistic, can be considered one of the most basic theories of physics, the choice introduced by the condition $p^0 \geq 0$ is transferred to the rest of physical theories. But such a choice is established from the very beginning, as a postulate of the theory. The challenge is, then, to *justify* the non time-reversal invariant postulate by means of independent theoretical arguments.

As it is well known, in the energy-momentum tensor $T^{0i}(x)$ represents the matter-energy flow and $T^{i0}(x)$ represents the linear momentum density. Since $T^{\mu\nu}$ is a symmetric tensor, $T^{\mu\nu}(x) = T^{\nu\mu}(x)$ and, therefore, $T^{0i}(x) = T^{i0}(x)$; this means that the matter-energy flow is equal to the linear momentum density. As a consequence, if the matter-energy flow $T^{0\alpha}$ can be used to define the arrow of time under the dominant energy condition, this is also the case for the linear momentum density $T^{\alpha 0}$. On the other hand, we have seen in Section VII that the local matter-energy flow $\tau^{\mu\nu}$ can be conceived as a conservative version of $T^{\mu\nu}$ in the orthonormal coordinates of a local inertial frame; in this case, the dominant energy condition has the consequence that $\tau^{0\mu}$ is non-spacelike. Now we know that exactly the same conclusion can be drawn for the local linear momentum density $\tau^{\mu 0}$. But it is precisely the local linear momentum density $\tau^{\mu 0}$ ($T^{0\mu}$ in the local inertial frame) the local magnitude corresponding to the classical p^μ of quantum field theory; thus, at each point x, $p^\mu \sim \tau^{0\mu} \in C_+(x)$.

In conclusion, the fact that, at each point x of the local context and, therefore, for every classical particle, p^μ must be contained in the future light semicone $C_+(x)$ turns out to be a consequence of the global time-asymmetry of the spacetime when the dominant energy condition holds everywhere. In other words, the non time-reversal invariant postulate can be justified on cosmological grounds instead of being imposed as a departing point of the axiomatic version of quantum field theory.

An analogous argument can be presented in the case of the ordinary version of quantum field theory. The classification of one-particle states according to their transformation under the Lorentz group leads to six classes of four-momenta (e.g. see [44], vol. I, page 65). However, it is considered that only three of these six classes have physical meaning: these are precisely the cases which agree with the non time-reversal invariant postulate of the axiomatic version of quantum field theory. In other words, the symmetry group of quantum field theory is the orthochronous group (see [28], [45]), where the space inversion P but not the time inversion T are included. This is another way of expressing the non time-reversal invariance of the quantum field theory. In this case, the non time-reversal invariance is introduced not by means of a postulate but on the basis of empirical arguments that make physically meaningless certain classes of four-momenta. However, to the extent that special relativity and standard quantum mechanics are time-reversal invariant theories, they give no theoretically grounded justification for such a breaking of time-reversal invariance.

Nevertheless, as we have seen in the previous subsection, this justification can be given on cosmological grounds.

Let us make the point in different terms. The quantum field correlates of P and T, $\mathbf{P}$ and $\mathbf{T}$, are defined as:

$$\mathbf{P} iP^{\nu}\mathbf{P}^{-1} = iP_{\mu}^{\nu}P^{\mu} \qquad \mathbf{T} iP^{\nu}\mathbf{T}^{-1} = iT_{\mu}^{\nu}P^{\mu} \tag{32}$$

where $\mathbf{P}$ is a linear and unitary operator and $\mathbf{T}$ is an antilinear and antiunitary operator. In fact, if $\mathbf{T}$ were linear and unitary, we could simply cancel the i's and, then, $\mathbf{T}P^{\nu}\mathbf{T}^{-1} = -P^{\mu}$: the action of the operator $\mathbf{T}$ on the operator P^{ν} would invert the sign of P^{μ}, with the consequence that the spectrum of the inverted energy-momentum operator would be contained in a past light semicone. In particular, for $\nu = 0$, $P^{\mu} = H$, where H is the energy operator; then, if $\mathbf{T}$ were linear and unitary, $\mathbf{T}H\mathbf{T}^{-1} = -H$ with the consequence that, for any state of energy E there would be another state of energy $-E$. The antilinearity and antiunitarity of $\mathbf{T}$ avoid these "anomalous" situations in agreement with the conditions imposed by the non time-reversal invariant postulate and, at the same time, make quantum field theory non time-reversal invariant. Once again, there are good empirical reasons for making $\mathbf{T}$ antilinear and antiunitary, but not theoretical justification for such a move.

Summing up, in quantum field theory it is always necessary to take a decision about the time direction of the spectrum of the energy-momentum operator P^{μ}. The point that we want to stress here is that, either in the case of the non time-reversal invariant postulate of the axiomatic version of quantum field theory or in the case of the usual version of quantum field theory, the decision can be justified on cosmological grounds, as a consequence of the global time-asymmetry of the universe and the dominant energy condition.

Finally, it is worth reflecting on the role of weak interactions in the problem of the arrow of time. The CPT theorem states that $\mathbf{CPT}$ is the only combination of charge conjugation $\mathbf{C}$, parity reflection $\mathbf{P}$ and time-reversal $\mathbf{T}$ which is a symmetry of quantum field theory. In fact, it is well known that weak interactions break the $\mathbf{T}$ of the CPT theorem. According to a common opinion, it is precisely this empirical fact the clue for the solution of the problem of the arrow of time: since the $\mathbf{T}$ symmetry is violated by weak interactions, they introduce a non-conventional distinction between the two directions of time (see [33]). The question is: is the breaking of $\mathbf{T}$ what distinguishes both directions of time in quantum field theory? As we have seen, the operator $\mathbf{T}$ was designed precisely to avoid that certain tetra-magnitudes, such as the linear momentum p^{μ}, have the "anomalous" feature of being contained in a past light semicone: the action of the operator $\mathbf{T}$ onto the energy-momentum operator P^{μ} preserves the time direction of P^{μ} and, therefore, of its eigenvalues. It is this fundamental fact what makes quantum field theory non time-reversal invariant, and not the incidental violation of $\mathbf{T}$ by weak interactions[14]. This non time-reversal invariance of quantum field theory, based on

---

[14] Of course, this leaves open a different problem: to explain why, among all the elementary interactions, only weak interactions break T.

the peculiar features of the operator $\mathbf{T}$, distinguishes by itself between the two directions of time, with no need of weak interactions. In other words, even if weak interactions did not exist, quantum field theory would be a non time-reversal invariant theory which would define the arrow of time. The real problem is, then, to justify the non time-reversal invariance of a theory which is presented as a synthesis of two time-reversal invariant theories such as special relativity and quantum mechanics. But this problem is completely independent of the existence of weak interactions and the breaking of $\mathbf{T}$ introduced by them. Summing up, weak interactions do not play a role as relevant in the problem of the arrow of time as it is usually supposed.

IX. Conclusions

The problem of the arrow of time is still at present one of the most controversial questions in the foundations of physics: many attempts to solve it have been made in the past with scarce consensus. Perhaps these poor results are the consequence of conceptual confusions that contaminate many interesting debates. For this reason, we have elucidated the central concepts involved in the discussion from the very beginning, and this task has allowed us to supply a clear and concise formulation of the problem. On the other hand, the emphasis on the difference between conventional and substantial distinctions in physics has prevented us from confusing the conventional use of temporally asymmetric labels like "past" and "future" with the substantial, theoretically grounded difference between the two directions of time.

In this paper we have argued for a global and non-entropic approach for addressing the problem of the arrow of time, according to which the arrow of time is a geometrical property of spacetime. In particular, we have tried to clearly state the conditions that a spacetime must satisfy for having a global and non-entropic arrow. But our main aim has been to show how a substantial time-orientation results from the geometrical time-asymmetry of spacetime, and how this global time-orientation is transferred to local contexts where the local theories of physics are valid. Finally, we have argued that the globally founded local arrow may serve for non-conventionally and non-pretheoretically deciding which one of the two time-symmetric twins resulting from local time-reversal invariant theories has physical meaning. This might help to remove many puzzles and circularities encountered in the traditional attempts to solve the problem of the arrow of time in terms of exclusively local arguments.

Acknowledgments

We are extremely grateful to Alejandro Gangui for his assistance regarding some astronomical subjects and to Arno Bhom for his kind hospitality at the University of Texas at Austin, where Section VIII was conceived. This paper was partially supported by CONICET, the Universidad de Buenos Aires, the Universidad Nacional de Quilmes and the Secretaría de Estado de Educación y Universidades of Spain.

REFERENCES

[7] M. Castagnino, F. Gaioli and E. Gunzig, *Found. Cosmic Phys.*, **16**, 221, 1996.
M. Castagnino and R. Laura, *Phys. Rev. A*, **56**, 108, 1997.
M. Castagnino, "The global nature of time asymmetry and the Bohm-Reichenbach diagram", in A. Bohm, H. Doebner and P. Kielarnowski (eds), *Irreversibility and Causality (Proc. G. 21 Goslar 1996)*, Springer-Verlag, Berlin, 1998.
M. Castagnino and E. Gunzig, *Int. Journ. Theo. Phys.*, **38**, 47, 1999.
M. Castagnino and C. Laciana, *Class. Quant. Grav.*, **19**, 2657, 2002.
M. Castagnino, J. Gueron and A. Ordoñez, *J. Math. Phys.*, **43**, 705, 2002.
M. Castagnino, G. Catren and R. Ferraro, *Class. Quant. Grav.*, **19**, 4729, 2002.
M. Castagnino, L. Lara and O. Lombardi, *Int. Jour. Theo. Phys.*, **42**, 2487, 2003.
M. Castagnino, O. Lombardi and L. Lara, *Found. Phys.*, **33**, 877, 2003.

[8] M. Castagnino, L. Lara and O. Lombardi, *Class. Quant. Grav.*, **20**, 369, 2003.

[9] A. Bohm, *Quantum Mechanics: Foundations and Applications*, Springer-Verlag, Berlin, 1979.

[10] A. Bohm and M. Gadella, *Dirac Kets, Gamow Vectors, and Gel'fand triplets*, Springer-Verlag, Berlin, 1989.

[11] C. G. Sudarshan, C. B. Chiu and V. Gorini, *Phys. Rev. D*, **18**, 2914, 1978.
M. Castagnino and E. Gunzig, *Int. Jour. Theo. Phys.*, **36**, 2545, 1997.

[12] A. Grünbaum, *Philosophical Problems of Space and Time*, Alfred A. Knopf, New York, 1963.

[13] J. J. Smart, "Time", in *Encyclopedia of Philosophy*, Macmillan, New York.

[14] L. Sklar, *Physics and Chance*, Cambridge University Press, Cambridge, 1993.

[15] R. Penrose, "Singularities and time asymmetry", in S. Hawking and W. Israel (eds.), *General Relativity, an Einstein Centenary Survey*, Cambridge University Press, Cambridge, 1979.

[16] R. G. Sachs, *The Physics of Time-Reversal*, University of Chicago Press, Chicago, 1987.

[17] H. Price, *Time's Arrow and the Archimedes' Point*, Oxford University Press, Oxford, 1996.

[18] P. Ehrenfest and T. Ehrenfest, *The Conceptual Foundations of the Statistical Approach in Mechanics*, Cornell University Press, Ithaca, 1959 (original 1912).

[19] S. Brush, *The Kind of Motion We Call Heat*, North Holland, Amsterdam, 1976.

[20] L. Boltzmann, *Annalen der Physik.*, **60**, 392, 1897.

[21] H. Reichenbach, *The Direction of Time*, University of California Press, Berkeley, 1956.

[22] P. C. Davies, *The Physics of Time Asymmetry*, University of California Press, Berkeley, 1974.

[23] P. C. Davies, "Stirring up trouble", in J. J. Halliwell, J. Perez-Mercader and W. H. Zurek (eds.), *Physical Origins of Time Asymmetry*, Cambridge University Press, Cambridge, 1994.

[24] R. P. Feynman, R. B. Leighton and M. Sands, *The Feynman Lectures on Physics, Vol. 1*, Addison-Wesley, New York, 1964.

[25] J. Earman, *Phil. Scien.*, **41**, 15, 1974.

[26] O. Penrose and I. C. Percival, *Proceed. of the Phys. Soc*, **79**, 605, 1962.

[27] G. Mattews, *Phil. Scien.*, **46**, 82, 1979.
[28] S. Schulman, *Phys. Rev. Lett.*, **83**, 5419, 1999.
[29] S. Hawking and J. Ellis, *The Large Scale Structure of Space-Time*, Cambridge University Press, Cambridge, 1973.
[30] B. F. Schutz, *Geometrical Methods of Mathematical Physics*, Cambridge University Press, Cambridge, 1980.
[31] S. F. Savitt, *Brit. Jour. Phil. Scie.*, **47**, 347, 1996.
[32] C. W. Misner, K. S. Thorne and J. A Wheeler, *Gravitation*, Freeman and Co., San Francisco, 1973.
[33] Y. Choquet-Bruhat and L. Lamoureau-Brousse, *C. R. Acad. Sci. Paris*, **276**, 1217, 1972.
P. Cordero and C. Teitelboim, *Ann. Phys.(N.Y.)*, **100**, 607, 1976.
T. Hughes, T. Kato and J. Marsden, *Arch. Ration. Mech. Anal.*, **63**, 273, 1977.
[34] A. Lichnerowicz, *J. Math. Phys.*, **17**, 2135, 1976.
[35] C. L. Bennet, M. Halpern, G. Hinshaw, N. Jarosik, A. Kogut, M. Limon, S. S. Meyer, L. Page, D. N. Spergel, G. S. Tucker, E. Wollack, E. L. Wright, C. Barnes, M. R. Greason, R. S. Hill, E. Komatsu, M. R. Nolta, N. Odegard, H. V. Peirs, L. Verde and J. L. Weiland, "First year Wilkinson Microwave Anisotropy Probe (WMAP)", *arXiv:astro-ph/0302207.*
A. Kogut, D. N. Spergel, C. Barnes, C. L. Bennet, M. Halpern, G. Hinshaw, N. Jarosik. M. Limon, S. S. Meyer, L. Page, G. Tucker, E. Wollack and E. L. Wright, "Wilkinson Microwave Anisotropy Probe (WMAP). First year", *arXiv:astro-ph/0302213.*
[36] S. Perlmutter and B. P. Schmidt, "Measuring cosmology with supernovae", *arXiv:astro-ph/0303428.*
[37] M. Castagnino, H. Giacomini and L. Lara, *Phys. Rev. D*, **61**, 107302, 2000.
M. Castagnino, H. Giacomini and L. Lara, *Phys. Rev. D*, **63**, 044003, 2001.
[38] M. Castagnino, *Phys. Rev. D*, **29**, 2216, 1989.
M. Castagnino and F. D. Mazzitelli, *Phys. Rev. D*, **42**, 482, 1990.
M. Castagnino and F. Lombardo, *Phys. Rev. D*, **48**, 1722, 1993.
[39] M. Visser, *Lorentzian Wormholes*, Springer-Verlag, Berlin, 1996.
[40] L. Landau and E. Lifchitz, *Théorie des Champs*, Editions Mir, Moscow, 1970.
[41] L. Sklar, *Space, Time and Spacetime*, University of California Press, Berkeley, 1974.
[42] P. D. Lax and R. S. Phillips, *Scattering Theory*, Academic Press, New York, 1979.
[43] A. Bohm, S. Maxson, M. Loewe, and M. Gadella, *Physica A*, **236**, 485, 1997.
[44] R. Laura and M. Castagnino, *Phys. Rev. A*, **57**, 4140, 1998.
R. Laura and M. Castagnino, *Phys. Rev. E*, **57**, 3948, 1998.
M. Castagnino and R. Laura, *Phys. Rev. A*, **56**, 108, 1997.
[45] A. Bohm, I. Antoniou and P. Kielanowski, *Phys. Lett. A*, **189**, 442, 1994.
[46] A. Bohm and S. Wickramasekara, Sujeewa, *Found. Phys.*, **27**, 969, 1997.
[47] R. Haag, *Local Quantum Physics. Fields, Particles, Algebras*, Springer, Berlin, 1996.
[48] N. Bogoliubov, A. A. Logunov and I. T. Todorov, *Axomatic Quantum Field Theory*, Benjamin-Cummings, Reading Mass., 1975.
[49] P. Roman, *Introduction to Quantum Field Theory*, Wiley, New York, 1969.
[50] S. Weinberg, *The Quantum Theory of Fields*, Cambridge University Press, Cambridge, 1995.
[51] P. Roman, *Elementary Theory of Particles*, North-Holland, Amsterdam, 1961.

In: Horizons in World Physics, Volume 248
Editor: A. Reimer, pp. 111-126
ISBN 1-59454-322-4

Chapter 5

LIGHT IN METRIC SPACE-TIME AND ITS DEFLECTION BY THE SCREW DISLOCATION

Miroslav Pardy [1]

Prague Asterix Laser System, PALS, Za Slovankou 3, 182 21 Prague 8, Czech Republic

and Department of Physical Electronics, and The Laboratory of the Plasma Physics,

Masaryk University, Kotlářská 2, 611 37 Brno, Czech Republic

ABSTRACT

We will explain the geometrical meaning of the metric of space-time. Also, we will discuss the problem of the dimensionality of space. We argue, by means of experimental facts, such as the molecule DNA, the galaxy formation, QED, Planck's law etc. that the external space is 3-dimensional. Its dimensionality cannot be proved a priori, in accordance with the Russell philosophy of mathematics. Then, we derive the light deflection caused by the screw dislocation in space-time. The derivation is based on the definition of the screw dislocation. Also on the idea that space-time is a medium which can be deformed in such a way that the deformation of space-time is equivalent to the existence of metric which is equivalent to gravity. The idea that space-time is a medium is tested by the gravitational time dilation in the nonrelativistic limit. The existence of the screw dislocation in the cosmology is hypothetically confirmed by observation of light or gamma cosmical bursts. This can be interpreted as the annihilation of the giant (screw) dislocations with the anti-dislocations where antidislocation is introduced in accordance with particle and string physics, where to every particle exists antiparticle and to every string exists antistring. The origin of gravitational bursts are analogical to the optical ones. The Hubble telescope is able to detect only the optical bursts. The gravitational bursts then can be detected by LIGO, VIRGO, GEO, TAMA and so on. We argue that cosmological constant plays a substantial role in the formation of the screw dislocations. The existence of dislocations and other topological defects in cosmology modify the distribution of the microwaves in cosmical background. The Misner-Wheeler geometrodynamics is considered as the alternative form of the topological defect origin of the elementary particles.

---

[1] E-mail: pamir@physics.muni.cz

Keywords: Metric, Deformation, Screw dislocation, Elementary particles.

INTRODUCTION

There is a possibility that during the big bang, supernova explosion, gravitational collapse, collisions of the high-energy elementary particles and so on, the dislocations in space-time were created. In this chapter we derive the light deflection caused by the screw dislocation in space-time.

In order to derive such deflection of light, it is necessary to explain the origin of metric in the Einstein theory of gravity.

Einstein gives no explanation of the origin of the metric, or, metrical tensor. He only introduces the Riemann geometry as the basis for the general relativity [1]. He "derived" the nonlinear equations for the metrical tensor [2] and never explained what is microscopical origin of the metric of space-time. Einstein supposed that it is adequate to the metric that it follows from differential equations as their solutions. However, the metric has a microscopical origin similarly to the situation where the phenomenological thermodynamics has also the microscopical and statistical origin.

Let us remember the different origins of metric. First, let us show that metric is generated by the coordinate transformations. We demonstrate it using the spherical transformations:

$$x = r\sin\theta\cos\varphi, \quad y = r\sin\theta\sin\varphi, \quad z = r\cos\theta. \tag{1}$$

The square of the infinitesimal element is as follows:

$$ds^2 = dx^2 + dy^2 + dz^2 = dr^2 + r^2 d\theta^2 + r^2 \sin^2\theta \, d\varphi^2. \tag{2}$$

We see that the nonzero components of the metrical tensor are

$$g_{11} = 1, \quad g_{22} = r^2, \quad g_{33} = r^2 \sin^2\theta. \tag{3}$$

For $r = const$, it is $dr = 0$ and the element of the length is the element of the 2-dimensional sphere in the 3-dimensional space. The resulting metric is not the 3-dimensional one but only 2-dimensional in the 3D space. So, in order to generate the 2D metric, it is necessary to use the 3D transformations in 3D space. If we want to generate the metric on the 3D sphere, then it is necessary to use the 4D transformations for x, y, z, ξ, where ξ is the extra-coordinate. So, the metric is generated by the curvilinear transformations. Einstein suggested the possibility that metric can be generated by gravitational field. He created the general theory of relativity and gravitation. Henri Poincaré never accepted the metric generated by the gravitational field.

The Riemann element $(ds)^2$ is defined as

$$ds^2 = g_{\mu\nu} dx^\mu dx^\nu \tag{4}$$

and it is composed from the four infinitesimal coordinate differentials. It means if we want to generate the metric of this 4-dimensional space-time by the coordinate transformations, then it is necessary to use the coordinate transformations in 5D space-time. In other words, it is necessary to introduce the extra-dimension. Einstein radically refused the extra-dimensions, he explained the curvature of space-time by introduction of the metric which depends on the temperature of the surface [3]. Of course such an explanation of the origin of metric is not generally accepted in the textbooks and monographs [4]. It was only a pedagogical explanation. Some mathematicians [5],[6] tried to prove that our space is 3-dimensional and they automatically excluded the extra-dimensions. However such proofs are misleading because we know from the Bertrand Russell philosophy of mathematics that the mathematical theorems are not existential. In other words, mathematics cannot say anything on the existence of electron, proton, quarks, strings and so on, because these things do not follow from the mathematical axiomatic system. They are things of the external world and not of the world of mathematics. At the same time pure mathematics cannot predict any fundamental physical constant, because every physical constant including also the fundamental ones is of the dynamical origin. Mathematics gives only such information which follow from its axiomatical systems.

Extra-dimensions can be introduced only by the definition and the existence of them cannot be mathematically proved. We know that the 3-dimensional space was confirmed by the most precise theory in the history of physics - QED, and it means that the extra-dimensions were not confirmed. Also the Planck law of the blackbody radiation in 4D space differs from the Planck law in 3D space. Similarly in genetics, the existence and the form of the molecule DNA can be considered as a proof of the 3-dimensionality of space. The formation of galaxies in 3D space substantially differs from the formation of galaxies in 4-dimensional space.

We know that the extra-dimensions can be compactificated. However, there is no physical law which enables compactification. Compactification is only the mathematical method of the string theory.

Einstein avoids the extra-dimensionality and compactification. He uses argumentation [3] on the existence of the noneuclidean geometry using the 2D hot plane, where the magnitude of a rule changes from point to point being dependent on the temperature at a given point. This method was also used by Feynman [7]. Rindler does not use this method of argumentation [4].

Deformation Origin of the Space-Time Metric

So, the question we ask, is, what is the microscopical origin of the metric of space-time. We postulate that the origin of metric is the specific deformation of the space-time continuum. We take the idea from the mechanics of continuum and we apply it to the space-time medium. A similar approach can be found in the Tartaglia article and his e-print [8], where space-time is considered as a deformable medium.

The mathematical description of the 3-dimensional deformation is given for instance in [9]. The fundamental quantity is the tensor of deformation expressed by the relative displacements u^i as follows:

$$u_{ik} = \left(\frac{\partial u_i}{\partial x^k} + \frac{\partial u_k}{\partial x^i} + \frac{\partial u^l}{\partial x^i} \frac{\partial u_l}{\partial x^k} \right); \quad i,k = 1,2,3. \tag{5}$$

The last definition can be generalized to the 4-dimensional situation by the following relation:

$$u_{\mu\nu} = \left(\frac{\partial u_\mu}{\partial x^\nu} + \frac{\partial u_\nu}{\partial x^\mu} + \frac{\partial u_\alpha}{\partial x^\mu} \frac{\partial u^\alpha}{\partial x^\nu} \right); \quad \mu,\nu = 0,1,2,3, \tag{6}$$

with $x^0 = ct, x^1 = x, x^2 = y, x^3 = z$.

In order to establish the connection between metric $g_{\mu\nu}$ and deformation expressed by the tensor of deformation, we write for the metrical tensor $g_{\mu\nu}$ of the squared space-time element

$$ds^2 = g_{\mu\nu} dx^\mu dx^\nu, \tag{7}$$

the following relation

$$g_{\mu\nu} = (\eta_{\mu\nu} + u_{\mu\nu}), \tag{8}$$

where

$$\eta_{\mu\nu} = \begin{pmatrix} 1 & 0 & 0 & 0 \\ 0 & -1 & 0 & 0 \\ 0 & 0 & -1 & 0 \\ 0 & 0 & 0 & -1 \end{pmatrix}. \tag{9}$$

Instead of work with the metrical tensor $g_{\mu\nu}$, we can work with the tensor of deformation $u_{\mu\nu}$ and we can consider the general theory of relativity as the 4-dimensional theory of some real deformable medium as a corresponding form of the metrical theory. First, let us test the deformation approach to the space-time in case of the nonrelativistic limit.

THE NONRELATIVISTIC TEST

The Lagrange function of a point particle with mass m moving in a potential φ is given by the following formula [10]:

$$L = -mc^2 + \frac{mv^2}{2} - m\varphi. \tag{10}$$

Then, for a corresponding action we have

$$S = \int L dt = -mc \int dt \left(c - \frac{v^2}{2c} + \frac{\varphi}{c} \right), \tag{11}$$

which ought to be compared with $S = -mc \int ds$. Then,

$$ds = \left(c - \frac{v^2}{2c} + \frac{\varphi}{c} \right) dt. \tag{12}$$

With $d\mathbf{x} = \mathbf{v}dt$ and neglecting higher derivative terms, we have

$$ds^2 = (c^2 + 2\varphi)dt^2 - d\mathbf{x}^2 = \left(1 + \frac{2\varphi}{c^2} \right) c^2 dt^2 - d\mathbf{x}^2. \tag{13}$$

The metric determined by this ds^2 can obviously be related to the u_α as follows:

$$g_{00} = 1 + 2\partial_0 u_0 + \partial_0 u^\alpha \partial_0 u_\alpha = 1 + \frac{2\varphi}{c^2}. \tag{14}$$

We can suppose that the time shift caused by the potential is small and therefore we can neglect the nonlinear term in the last equation. Then we have

$$g_{00} = 1 + 2\partial_0 u_0 = 1 + \frac{2\varphi}{c^2}. \tag{15}$$

The elementary consequence of the last equation is

$$\partial_0 u_0 = \frac{\partial u_0}{\partial (ct)} = \frac{\varphi}{c^2}, \tag{16}$$

or,

$$u_0 = \frac{\varphi}{c} t + const. \tag{17}$$

Using $u_0 = g_{00} u^0$, or, $u^0 = g_{00}^{-1} u_0 = \frac{\varphi}{c} t$, we get with $const. = 0$ and

$$u^0 = ct' - ct, \tag{18}$$

the following result

$$t'(\varphi) = t(0) \left(1 + \frac{\varphi}{c^2} \right), \tag{19}$$

which is the Einstein formula relating time in the zero gravitational field to time in the gravitational potential φ. The time interval $t(0)$ measured remotely is so called the coordinate time and $t(\varphi)$ is local proper time. The remote observer measures time intervals to be deleted and light to be red shifted. The shift of light frequency corresponding to the gravitational potential is, as follows [10].

$$\omega = \omega_0\left(1+\frac{\varphi}{c^2}\right). \tag{20}$$

The precise measurement of the gravitational spectral shift was made by Pound and Rebka in 1960. They predicted spectral shift $\Delta\nu/\nu = 2.46\times10^{-15}$ [1]. The situation with the red shift is in fact the closed problem and no additional measurement is necessary.

While we have seen that the red shift follows from our approach immediately, without application of the Einstein equations, it is evident that the metric determined by the Einstein equations can be expressed by the tensor of deformation. And vice versa, to every tensor of deformation the metrical tensor corresponds.

The Deflection of Light by the Screw Dislocation

The problem of light deflection by the screw dislocation is a problem of the recent years [11], [12], [13], [14], [15], and so on. The motivation was the old problem of the deflection of light by the gravitational field which according to Einstein causes the curvature of space-time.

We know from the history of physics that the deflection of light by the gravitational field of the Sun was first calculated by Henri Cavendish in 1784, it was never published. The first published calculation was almost 200 years later in 1981 by the Prussian astronomer Johann Soldner. Einstein's calculation in 1911 was 0.83 seconds of arc. Cavednish and Soldner predicted a deflection 0.875 seconds of arc. So, the prediction of Cavendish, Soldner [16] and Einstein in 1911 were approximately half of the correct value which was derived in 1919 by Einstein.

Einstein in 1911 used the principle of equivalence for the determination of the light deflection. As was shown by Ferraro [17] the Einstein application of this principle was incorrect. The correct application was given only by Ferraro in order to get the correct value. The deflection of light by the topological defects as dislocations, disclinations and so on was, to my knowledge, never calculated by Einstein. In recent times, such calculations were performed by [11], [12], [13] [14], [15] and so on. Here we use a different and simpler method, a definition of the screw dislocations that differs from the above authors.

According to [9], the screw deformation in the mechanics of continuum was defined by the tensor of deformation which is in the cylindrical coordinates as

$$u_{z\varphi} = \frac{b}{4\pi r}, \tag{21}$$

where b is the z-component of the Burgers vector. The Burgers vector of the screw dislocation has components $b_x = b_y = 0, b_z = b$. The Burgers vector is for the specific dislocation a constant geometrical parameter.

The postulation of the space-time as a medium enables the transfer of notions to the theory of elasticity into the relativistic theory of space-time and gravity. The considered transfer is of course the heuristical operation, nevertheless the consequences are interesting. To our knowledge, the problem, which we solve is new.

We know that the metric of the empty space-time is defined by the coefficients in the relation:

$$ds^2 = c^2dt^2 - dr^2 - r^2d\varphi^2 - dz^2. \tag{22}$$

If the screw deformation is present in space-time, then the generalized metric is of the form:

$$ds^2 = c^2dt^2 - dr^2 - r^2d\varphi^2 - 2u_{z\varphi}dzd\varphi - dz^2, \tag{23}$$

or,

$$ds^2 = c^2dt^2 - dr^2 - r^2d\varphi^2 - \frac{2b}{4\pi r}dzd\varphi - dz^2. \tag{24}$$

The motion of light in the Riemann space-time is described by the equation $ds = 0$. It means, that from the last equation the following differential equation for photon follows:

$$0 = c^2 - \dot{r}^2 - r^2\dot{\varphi}^2 - \frac{b}{2\pi r}\dot{z}\dot{\varphi} - \dot{z}^2. \tag{25}$$

Every parametric equation which obeys the last equation are equations of motion of photon in the space-time with the screw dislocation. Let us suppose that the motion of light is in the direction of the z-axis. Or, we write approximately:

$$r \approx a; \quad \dot{z} = v. \tag{26}$$

Then, we get equation of φ:

$$2\pi a^3\dot{\varphi}^2 + bv\dot{\varphi} = 2\pi a(c^2 - v^2). \tag{27}$$

We suppose that the solution of the last equation is of the form

$$\varphi = At. \tag{28}$$

Then, we get for the constant A the quadratic equation

$$2\pi a^3A^2 + bvA + 2\pi a(v^2 - c^2) = 0 \tag{29}$$

with the solution

$$A_{1/2} = \frac{-bv \pm \sqrt{b^2 v^2 - 16\pi^2 a^4 (v^2 - c^2)}}{4\pi a^3}. \tag{30}$$

Using approximation $v \approx c$, we get that first root is approximately zero and for the second root we get:

$$A \approx \frac{-bc}{2\pi a^3}, \tag{31}$$

which gives the function φ in the form:

$$\varphi \approx \frac{-bc}{2\pi a^3} t. \tag{32}$$

Then, if $z_2 - z_1 = l$ is a distance between two points on the straight line parallel with the axis of screw dislocation then, $\Delta t = l/c$, c being the velocity of light. For the deflection angle $\Delta\varphi$, we get:

$$\Delta\varphi \approx \frac{-bl}{2\pi a^3}. \tag{33}$$

So, we can say, that if we define the screw dislocation by the metric of eq. (24), then, the deflection angle of light caused by such dislocation is given by eq. (33). The result (33) is only approximative and we do not know what is the accuracy of such approximation. This problem can be solved using the approximation theory.

Let us remark that the exact trajectory of photon in the field of the screw dislocation can be determined from the trajectory equation

$$\frac{d^2 x_\mu}{ds^2} + \Gamma_\mu^{\alpha\beta} \frac{dx_\alpha}{ds} \frac{dx_\beta}{ds} = 0. \tag{34}$$

which was used in many textbooks. However, According to Landau et al. [10], the equation is contradictory for photon, because in this case $ds = 0$, and it means that the last equation is not rigorously defined. Landau et al. derived the deflection of light from the Hamilton-Jacobi equation for particle with the rest mass $m = 0$, which moves with the light velocity. However, this approach is not also absolutely correct because in the classical field theory it is not possible to define photon. Photon is a quantum object. Rigorous derivation of the deflection of light was given by Fok [18], who used the mathematical object "the front of wave" and his result is valid without any doubt.

Let us remark that equation (34) has two meanings: geometrical and physical. The geometrical meaning uses $g_{\mu\nu}$ which follows from the curvilinear transformations and the physical meaning of $g_{\mu\nu}$ is metric of the gravitational field calculated by means of the Einstein equation. The second meaning is the Einstein postulate and cannot be derived from so called pure mathematics. Only experiment can verify the physical meaning of equation (34).

The problem of interaction of light with the gravitational field is not exhausted by our example. We can define more difficult problems such as deflection of the coherent light, laser light, squeezed light, soliton light, massive light with massive photons, light of the entangled photons and so on. None of these problems has been solved because they are only for brilliant experts very well educated. And this is the pedagogical problem.

THE PHYSICAL GENERATION OF THE SCREW DISLOCATION OF THE SPACE-TIME

Now the question arises, how to determine the mechanical or electrodynamical or laser system which will generate the screw dislocation in space-time. We know, that for real crystals the generation of the screw dislocation is the elementary problem of the physics of crystals. If we use Einstein's equations, then the problem is mathematical one. Or, to see it, let us write the Einstein equations with the cosmological constant λ.

$$R_{\mu\nu} - \frac{1}{2} g_{\mu\nu} R + \lambda g_{\mu\nu} = -\kappa T_{\mu\nu}, \tag{35}$$

where $T_{\mu\nu}$ is the tensor of the energy and momentum.

In case of the perfect fluid and pressure, tensor of the energy and momentum is as follows:

$$T_{\mu\nu}(\text{mech}) = (\varrho + p)u_\mu u_\nu + p g_{\mu\nu}, \tag{36}$$

where ϱ is a density and p is a pressure of the fluid. The quantities u_μ are four velocities of the fluid.

In case that the tensor of the energy and momentum is created electromagnetically, then,

$$T_{\mu\nu}(\text{elmag}) = \frac{1}{4\pi}\left(F_{\mu\alpha} F_\nu^\alpha - \frac{g_{\mu\nu}}{4} F^{\alpha\beta} F_{\alpha\beta} \right). \tag{37}$$

where $F_{\alpha\beta}$ is the tensor of the electromagnetic field.

So, because $g_{\mu\nu}$ is determined as the metrical tensor corresponding to the screw dislocation of the space-time, the left side of the Einstein equations is given and the problem is to find the mechanical and electrodynamical quantities, which determine corresponding tensors of energy and momentum.

The surprising thing is the fact that if there is no curvature of space time, then thanks to the existence of the cosmological constant, the solution corresponding to the mechanical or electrodynamical systems is not absolutely zero. The cosmological constant is a very important quantity and evidently cannot be zero. The dislocations of space-time are in harmony with the cosmological constant.

Let us remark that Einsteins equations were derived intuitively by Einstein [2] and rigorously by Hilbert from the Lagrangian using the variational method [1]. The Hilbert derivation is a pure mathematical one which means it is very simple. This variational method enables to start relativity physics from the general theory and to derive the special relativity as a classical limit of the general theory. This approach was presented by Rindler [19]. To our knowledge, such unconventional but very elegant approach was not presented in any textbook on relativity theory.

The tensor of the energy momentum in equation (36) is rigorously defined. The problem is, how to identify the distribution of the cosmological objects with this tensor. To our knowledge, there is no mathematical theorem for such rigorous identification.

Cosmological Consequences

The verification of our theory and at the same time the existence of the nonzero cosmological constant can be performed during the measurement of the cosmical microwave background radiation. In case of the existence of some defects in space-time the distributions of this radiation will be inhomogeneous and it will depend on the density and orientations of the dislocations, screw dislocations, disclinations and other topological defects of the space-time.

It is also evident in case of the curvature caused by some topological defect is zero, then, thanks to the existence of the cosmological constant in the Einstein equations the defects can be generated mechanically or electrodynamically as it follows from the Einstein equations. So, investigation of the cosmical microwave background can inform us on the distribution of the topological defects in the space-time and on the possible origins of these defects [20].

The Laboratory Verification of the Theory

We can consider the situation which is analogical to the space-time situation. In other words, we can consider the modified Planckian experiment with the black body radiation. The difference from the original Planck situation is that we consider inside of the black body some optical medium with dislocations. Then, in case that the optical properties depend also on the presence of dislocations, in other words, that the local index of refraction depend on presence of dislocations, then we can expect the modification of the Planck law of the blackbody radiation. We know that the most simple problem with the constant index of refraction was calculated [21]. To our knowledge, although this is only a so called table experiment, it was never performed in an optical laboratory. It is possible to expect that during the experiments some surprises will appear. However, the practical situation can be realized, if we prepare some crystal with the screw dislocations with given orientation. Then, in case that the optical properties are expressed by means of metric in crystal, the metric will determine the optical path of light in the crystal and it means that the screw dislocations can be investigated by optical methods and not only by the electron microscope.

DISCUSSION

We have defined in harmony with the author article [22], [23] gravitation as a deformation of a medium called space-time. We have used an equation which relates Riemann metrical tensor to the tensor of deformation of the space-time medium and applied it to the gravitating system, which we call screw dislocation in space-time. The term screw dislocation was used as an analogue with the situation in the continuous mechanics. We derived the angle of deflection of light passing along the screw dislocation axis at the distance a from it on the assumption that trajectory length was l. This problem was not considered for instance in the Will monograph [24]. The screw dislocation was still not observed in space-time and it is not clear what role the dislocations plays in the development of the universe after the big bang. Our method can be applied to the other types of dislocations in space-time and it is not a problem to solve this problem in general. We have used here the specific situation because of its simplicity. We have seen that the problem of dislocation in space-time is interesting and it means there is some scientific value to this problem.

It is well known from the quantum field theory and experiment, that every particle has its partner in the form of the antiparticle [25]. For instance the antiparticle to the electron is positron. It is well known that after annihilation of particle-antiparticle pair, photons are generated. For instance

$$e^{+} + e^{-} \quad \rightarrow \quad 2\gamma. \tag{38}$$

Now, if we define that to every dislocation D exists antidislocation $\bar{D}$, (which can be considered also as an analogue of the antistring, [26]), then, we can write the following equation which is analogical to (38)

$$D + \bar{D} \quad \rightarrow \quad N\gamma \tag{39}$$

where N is natural number and gamma denotes photon. The next equation is also possible:

$$D + \bar{D} \quad \rightarrow \quad Ng \tag{40}$$

where g denotes graviton.

It is possible also to consider the high-energy process with the incident particles a and b as follows:

$$a + b \quad \rightarrow \quad c_1 + c_2 + c_3 + \ldots c_n, \tag{41}$$

where c_i are denotations of identical or different particles. It is well-known that the equation (41) will be the fundamental equation of LHC.

In case of the existence of the dislocations in universe, equations (39) and (40) represents the burst of photons or gravitons in the cosmical space. So, we defined the further possible interpretation of the photonic and gravitational bursts in cosmical space.

Let us remark that the dislocation approach to the particle physics are in harmony with the Einstein dream and later Misner-Wheeler geometrodynamics where all existing elementary objects can be defined as some form of space-time. Misner and Wheeler [27], [28] also consider that neutrino is the specific form of space-time. Let us still remark that we know from the history of philosophy that long before Christ, Anaximandros introduced **apeiron** as a medium from which all particles, and therefore all visible universe was created. So, we can say that the famous trinity of men, Einstein-Misner-Wheeler, is the follower of Anaximandros.

The identification of the fundamental particles by the dislocations is in harmony with the relation for the energy of the dislocation [29]

$$E = \frac{E_0}{\sqrt{1 - \frac{v^2}{c^2}}}, \tag{42}$$

where E_0 is the energy of dislocation when its velocity v is zero and c is the velocity of sound in the crystal. In case of medium called space-time the velocity c is the velocity of light in vacuum. So, vacuum is in a certain sense medium which is similar to the crystal medium. The analogy with the dislocations in crystal is of course heuristical step which is the integral part of the methodology of discoveries and it cannot be rigorously algorithmically defined.

Equation (39) can be verified by the table experiment, where the dislocations and the anti-dislocations in some crystal are annihilated by some suitable laboratory mechanism. The flash of light is observed. The laboratory mechanism can be realized for instance by the ultrasound, pressure, strike, and so on. The application of the ultrasound is analogical to the case of the sonoluminescence where the mechanism of the generation of light involves application of ultrasound. We hope that the energy released by such annihilation is greater than the energy released by the mechanism of cold fusion.

Although equation (42) can be identified with the relativistic equation for dependence of energy on velocity of elementary particle, we cannot identify electron with the screw dislocation in space-time. Why? Because we know that the attractive or repulsive force between two screw dislocations are different than the force between two electrons, two positrons or electron and positron. The second reason is that the anomalous magnetic moment of electron is of the dynamical origin as it follows from the Feynman diagram technique while the classical dislocation does not involve such dynamics. On the other hand, we do not know how other dislocations such as circular dislocations, cylindrical dislocations, helix, double helix, triple helix dislocations and so on are related to elementary particles, especially to neutrinos. We know that all physical constant in the standard model are of the dynamical origin, but at present time it is not clear what is their derivation from the more fundamental theory (subquark theory, string theory and so on), or, from the dislocation theory of elementary particles. We think that dislocation theory of elementary particles is not at present time prepared to give the answer to these difficult questions.

The 3D screw dislocation can be extended mathematically to the N-dimensional space, or, space-time. However the interpretation of the N-dimensional theory needs introducing of the compactification.

So, we can say that the Einstein dream of the unification of all objects and forces in nature in the framework of geometrodynamics is far from a successful realization because the identification of ultimate blocks of nature with dislocations and with the topological defects is not elaborated at this time.

According to Veltman, there are plenty of mysteries in particle physics [30]. We also know one mystery in mathematics. This is the imaginary number $i = \sqrt{-1}$. It is considered usually as the mysterious number because there is no geometrical meaning of this number. The mathematical relation

$$i^i = e^{-\pi/2} \tag{43}$$

is not mysterious, because the proof of this relation is elementary. In physics the situation is such, that if we do not know what the physical meaning is of some relation, then it is mysterious.

We know that Parmenides was not able to understand motion. Motion was mysterious for him. Why? Because he was not able to introduce time in his system of thinking. In particle physics and in the string theory [31] the confinement of quarks is mysterious and the solution of his problem in the dislocation theory of fundamental constituents is open. We think, that appropriate understanding of the string-like dislocations [32] and definition of the ultimate building blocks of nature can solve all problems of particle physics together with removing all mysteries.

While the verification of the existence of the optical bursts caused by the annihilations of the giant screw dislocations and anti-dislocations can be detected by the Hubble telescope, the gravitational bursts can be probably detected by LIGO [33], VIRGO [34], GEO [35], TAMA [36], and so on.

Let us remark that the observations of recent gamma-ray bursts GRB021004 and GRB990123 are interpreted as arising from the ultrarelativistic particles thrown out by a cataclysmic event as the collapse of massive star in the so called "cannonball" model. Such models apply to gamma-ray bursts lasting several seconds. Bursts shorter than about 2 seconds are thought to be due to coalescence of two neutron stars to form a black hole. So, our interpretation of light bursts differs from the traditional one because we explain the light bursts as an annihilation of dislocations and anti-dislocations existing in space-time. We hope that our interpretation of gamma-bursts based on particle physics, topology and general relativity is progressive.

REFERENCES

[52] I. R. Kenyon, *General Relativity*, (Oxford University Press Inc., New York, 1996).

[53] S. Chandrasekhar, On the "Derivation of Einstein's Field Equations", Amer. Journal of Phys., No. 2, (1972), p. 224.

[54] A. Einstein, *Über die specielle und die allgemaine Relativitätstheorie, Gemeinverstandlich*, (Vierte Auflage, Verlag Vieweg & Sohn, Braunschweig, 1919).

[55] 4 W. Rindler, *Relativity, Special, General and Cosmological*, (Oxford University Press Inc., New York, 2003).

[56] P. Natorp, Zu den logischen Grundlagen der neueren Mathematik, Archiv für systematische Philosophie, 1901, p. 189.

[57] P. Natorp, *Die Logischen Grundlagen der exakten Wissenschaften*, p. 189, (1901).

[58] R. P. Feynman, F. B. Morinigo and W. G. Wagner, *Feynman Lectures on Gravitation*, (Penguin Books Ltd, 27 Wrights Lane, London W8 5TZ, England 1999).
[59] A. Tartaglia, Four Dimensional Elasticity and General Relativity, Grav. Cosmol., (1995), p. 335; e-print gr-qc/9509043.
[60] L. Landau, E. Lifshitz, *Theory of Elasticity* (Butterworth-Heinem, New York, 1995).
[61] L. Landau, E. Lifshitz, *The Classical Theory of Fields*, (Pergamon Press, Oxford, 1962).
[62] M. O. Katanaev and I. V. Volovich, Scattering on Dislocations and Cosmic Strings in the Theory of Defects, e-print gr-qc/9801081.
[63] M. O. Katanaev and I. V. Volovich, Theory of Defects in Solids and Three-Dimensional Gravity, Annals of Physics (NY), (1992), p. 1.
[64] F. Moraes, Geodesics Around a Dislocation. Phys. Lett. A, (1996), p. 189.
[65] A. de Padua, F. Parisio-Filho and F. Moraes, Geodesics Around Line Defects in Elastic Solids, Phys. Lett. A, (1998), p. 153.
[66] L. C. Garcia de Andrade, Geodesics Around Gravitational Dislocations, e-print gr-qc/9811018.
[67] P. M. Brown, Einstein's Gravitational Field, e-print physics/0204044.
[68] R. Ferraro, The Equivalence Principle and the Bending of Light, Am. J. Phys. (2), February (2003), p. 168.
[69] V. A. Fok, *The Theory of Space, Time and Gravity*, (GIF, Moscow, 1961), (in Russian).
[70] W. Rindler, General Relativity before Special Relativity: An Unconventional Overview of Relativity Theory, Am. J. Phys., (10), October (1994), p. 887.
[71] M. Lachièze-Rey and E. Gunzig, *The Cosmological Background Radiation*, (Cambridge University Press, 1999).
[72] Ryogo Kubo, *Statistical Mechanics*, (North Holland Publishing Company, 1965).
[73] M. Pardy, Renormalization of Gravity and Gravitational Waves, gr-qc/0103010.
[74] M. Pardy, Deflection of Light by the Screw Dislocation in Space-Time, e-print gr-qc/0106019.
[75] C. M. Will, *Theory and Experiment in Gravitational Physics*, Revised edition, (Cambridge University Press, Cambridge, 1983).
[76] L. Maiani and R. A. Ricci, ed., Conference Proceedings, Symposium in Honour of Antonio Zichichi to Celebrate the 30th Anniversary of *The Discovery of Nuclear Antimatter*, (Bologna, 18 December 1995).
[77] A. M. Srivastava, Numerical Study of the Lowest Energy Configurations for Global String-Antistring Pairs, e-print hep-ph/9209281.
[78] C. W. Misner and J. Wheeler, Classical Physics as Geometry: Gravitation, Electromagnetism, Unquantized Charge and Mass as Properties of Curved Empty Space, Ann. Phys. (NY)No. 6, (1957), p. 525.
[79] J. Wheeler, On the Nature of Quantum Geometrodynamics, Ann. Phys. (NY)No. 6, (1957), p. 604.
[80] A. H. Cottrell, *Theory of Crystal Dislocations*, (Blackie and Son Ltd., London, Glasgow 1964).
[81] M. Veltman, *Facts and Mysteries in Elementary Particle Physics* (World Scientific Publishing Company, Pte. Ltd., Singapore 596224, 2003).
[82] I. Antoniadis, Testing Times for Strings, CERN Courier, July/August (2003).

[83] F. Lund, Equation of Motion of a Stringlike Dislocation, Phys. Rev. Lett., No. 1, (1985), p. 14.

[84] B. C. Barish, "LIGO: Status and Prospects", Proc. of Conference on Gravitational Wave Detection, Tokyo, Eds. K. Tsubono, M. K. Fujimoto, K. Kuroda, p. 163, (Univ. Academy Press Inc., Tokyo, 1997).

[85] J. -Y. Vinet et al., Proc. of Gravitation and Cosmology, ICGC-95 Conference, Pune, India 13-19 Dec. 1995, Astrophysics and Space Science Library, p. 89, (Kluver Academic Publishers, 1997).

[86] J. Hough et al., "GEO 600: Current Status and Some Aspects of the Design" Proc. of Conference on Gravitational Wave Detection, Tokyo, Eds. K. Tsubono, M. K. Fujimoto, K. Kuroda, p. 175, (Univ. Academy Press Inc., Tokyo, 1997).

[87] K. Tsubono and TAMA collaboration, "TAMA project" Proc. of Conference on Gravitational Wave Detection, Tokyo, Eds. K. Tsubono, M. K. Fujimoto, K. Kuroda, p. 183, (Univ. Academy Press Inc., Tokyo, 1997).

In: Horizons in World Physics, Volume 248
Editor: A. Reimer, pp. 127-166

ISBN 1-59454-322-4

Chapter 6

MOTION OF EXTENDED BODY ON CURVED SPACETIME

Akira Ohashi

Department of Physics, Ochanomizu University, 2-1-1,

Otsuka, Bunkyo, Tokyo, 112-8610, Japan

ABSTRACT

In this chapter, we will develop a *multipole particle* approach to describe the motion of extended bodies on a curved spacetime. The concept of the multipole particle is to replace extended bodies with point-like particles and to describe motion of the objects by the particles. The particle can have arbitrary multipole structures and its multipole moments are defined by the gravitational field created by the particle. We construct the energy momentum tensor for such a particle. Once the energy momentum tensor is constructed, we can derive equations of motion for the particle. As an example, we will derive relativistic equations of motion for spinning particles, and discuss the practical applications to the post-Newtonian approximation of Einstein's theory of gravitation.

INTRODUCTION

"Motion"– this is perhaps the most familiar phenomenon to us. Maybe, this is the first physical phenomenon that human beings have considered with great curiosity. It is also the most important physical phenomenon for us. In the seventeenth century, Newton discovered the mathematical description for the motion of objects. After this discovery, our scientific knowledge literally exploded. Today, this theory by Newton is known as *dynamics* or *mechanics*. It scarcely needs to be said that mechanics is the most important foundation to modern physics, science, engineering and so on.

Newton also discovered the law of universal gravitation. Maybe, people know this amusing story: he discovered gravitation by seeing an apple dropping from a tree. If we interpret this anecdote, he found the existence of a force that must act on objects at the sight of a falling object; there obviously a relation existed between the force and the motion as a background of this discovery. Now, Newton's theories have achieved great success. It is because his theory can describe not only the motion of familiar objects to us, but also the motion of the celestial objects. Although we can not touch and verify them directly, Newton's theories can describe and predict their motion very accurately.

Now, let us consider the following three laws of motion, known as Newton mechanics,

1. *Every body perseveres in its state of rest, or uniform motion in a straight line, except in so far as it is compelled to change that state by forces impressed on it* (this law is regarded as a definition of inertial systems and a definition of force.)
2. *The rate of change of linear momentum is proportional to the force applied, and takes place in the straight line in which that force acts* (this explains the equation of motion.)
3. *Forces are caused by the interactions of pairs of bodies. The force exerted by A upon B and the force exerted by B upon A are: simultaneous; equal in magnitude; opposite in direction; in the same straight line; caused by the same mechanism* (this is known as the law of action and reaction.)

The statements in italic letters are quoted from Penguin Books (2000) [1].

The author's comments for each laws are presented in the parenthesis. These fundamental laws of mechanics can describe all motion of all objects. However, in constructing the theory of mechanics, we usually apply these laws to an ideal object, *point mass*. Then, we integrate the equation of motion over the extent of an object and obtain the equation of motion that can describe the motion of the extended object. The linearity of the equation of motion to the mass allows this integration. Thus, we get a theory of the motion of rigid bodies. In turn, when we apply the equation of motion of point masses to infinitesimal elements of objects, we can construct so-called fluid mechanics.

The same argument of integration is possible for the law of universal gravitation, because the gravity is linear to the mass of a point mass that generates the gravity. As well as the above argument for rigid bodies, we can obtain the gravitational force by the object by integrating the law of gravitation over the extent of the object. Then, what kinds of questions are expected when we combine the laws of motion and the law of gravitation?

First, when we consider a lump of matter, we can discuss a problem of the structure of the lump, for example, a *star*. The structure of stars is one of the most important themes of astrophysics. We can obtain much knowledge of the structure of stars by combining the law of gravitation and fluid mechanics (we need some knowledge about the matter, *equation of states etc.*, to get sensible results) [2]. This is the first problem one should tackle.

Next, let us consider more than two lumps of matter. Of course, we can discuss a subject of the structure of each *star*. However, when there are plural lumps, another question arises: how does the one object move against the others. Otherwise, we can also argue a question of motion of the extended body in the external gravitational field. The simplest situation of this type is that all stars are spherically symmetric. In this case, we can replace the lumps of matter with point masses and describe the motion of the stars. This replacement is possible when the stars are sufficiently distant from each other and any tidal forces can be neglected, for example, such as the motion of the planets around the sun. In such a situation, the knowledge for point masses is enough.

Nevertheless, the point mass approximation becomes imperfect, when we discuss motion of close binaries with short periods or motion of satellite near large planets such as Jupiter. The insufficiency is also realized when we discuss accurate motion over very long duration even if bodies are distant from each other. In these cases, we must consider the distribution of the matter and put it in the equation of motion. Usually, uneven distribution of matter is expressed by multipole moments, and then the equation of motion has terms by these moments. In other words, we must consider not only the motion of *stars* but also the structure of the *stars* at the same time. This is a very interesting problem and some interesting results are known in certain ideal situations [3].

There are important problems, such as above, related only to Newton's laws of motion and law of universal gravitation in astronomy and astrophysics. Furthermore, at present, we know that gravitation must be described by general relativity more accurately. Then, the same types of problems are also considered to be important in the situation where relativity is needed as well as Newtonian situation. We, ourselves, rather find the problems of motion in gravitational fields as a more important problem. Accordingly, the theme of this chapter is to describe the motion of objects.

The important difference between Newton's law of universal gravitation and Einstein's general relativity is that the gravity described by general relativity is a dynamical object but Newtonian gravity is not. This means that there exist gravitational waves caused by the dynamical property of gravitation in general relativity. The existence of the gravitational waves was expected soon after Einstein had proposed his theory of gravitation. However, it still took sometime before people began to pay attention to the gravitational waves as a realistic phenomenon. The cue is a discovery of a pulsar in a binary system by R. A. Hulse and J. H. Taylor [4]. The binary pulsar discovered by them provides a very important tool to test the existence of the gravitational waves. A curious observation: the period of the orbital motion of the binary system changes at the same rate with that calculated by general relativity. The binary loses their gravitational energy by radiating the gravitational waves and the orbital separation decreases. Thus, the period changes. Namely, these observation confirm the existence of the gravitational waves indirectly.

Laser interferometric gravitational wave detectors are under construction at some places in order to detect gravitational waves directly and to open a door to gravitational wave astronomy. These detectors are intended to observe mainly binary systems that consist of compact objects such as neutron stars and so on. Among these binaries, they aim at moments that the objects of the binary get closer by emitting gravitational waves and are about to coalesce each other into a new object. The reason why these phenomenon can be targets of the detectors is that this coalescing can emit gravitational waves and the physics of this phenomenon is comparatively clear. The former reason is due to possibilities of observing distant events. The latter reason is from a request of method for the data analysis. The method adopted to analyze the datum is the matched filtering method. This method uses a theoretically calculated template of the gravitational wave forms in order to cut noises from the raw data outputted by the detectors.

The problem of generation of gravitational waves is an important and realistic motivation to study celestial motions. In the matched filtering method, many templates are calculated in advance and are correlated to the raw data from the detectors. Then, we find a template that makes the S/N ratio the largest. Thus, we can obtain physical quantities of the target by the parameters of the selected template. Consequently, in order to get more precise parameters, we need more accurate templates. In these calculations, it is important how an extended body moves in strong gravitational field. This problem makes studies of relativistic motion of objects more important. Our main theme in this chapter is to develop the multipole particle description as an approach to solve the motion of extended bodies in relativistic gravitational fields [5].

In section 2, we will briefly review the history of extended bodies. In section 3, we explain the key idea of our theory by using an example of the electric field. In section 4, we apply the idea in section 3 to a linearized gravitational field. In section 5, we discuss the problem arisen by simultaneity in a spacetime. Then, we will introduce *multipole particles* in the special relativity in section 6, and in the general relativity in section 7. In section 8, we will apply our theory to some cases. Finally, we will refer to some prospects of our description for motion of objects and some matters left over, then summarize our story.

Notations used in this chapter are the following. The signature of metrics is $(-,+,+,+)$. We denote metrics of a Minkowski spacetime and a curved spacetime by $\eta_{\mu\nu}$ and $g_{\mu\nu}$, respectively. The Roman indices run over 1 to 3 and the Greek indices run over 0 to 3. The definition of curvature is $R^{\mu}{}_{\nu\alpha\beta} = \Gamma^{\mu}{}_{\nu\beta,\alpha} - \Gamma^{\mu}{}_{\nu\alpha,\beta} + \Gamma^{\mu}{}_{\alpha\sigma}\Gamma^{\sigma}{}_{\nu\beta} - \Gamma^{\mu}{}_{\beta\sigma}\Gamma^{\sigma}{}_{\nu\alpha}$ and $R_{\mu\nu} = R^{\sigma}{}_{\mu\sigma\nu}$. We use convenient symbols $\partial_{\alpha\beta\cdots\gamma} = \partial_{\alpha}\partial_{\gamma}\cdots\partial_{\gamma}$ and $\nabla_{\alpha\beta\cdots\gamma} = \nabla_{\alpha}\nabla_{\beta}\cdots\nabla_{\gamma}$ for simplicity of writing. We occasionally use the multi-index notation for capital Roman indices; in the expression $A_L B^L$, *the upper-case letter* "L" indicates an ordered set of *lower-case letter* "l" *indices*, i.e., "l" is a mere number of the indices, i.e. $A_L B^L = A_{k_1\cdots k_l} B^{k_1\cdots k_l}$. The signature of perfect asymmetry symbol ε's is defined by $\varepsilon^{123} = \varepsilon_{123} = +1$ for 3-dimensional and $\varepsilon^{0123} = -\varepsilon_{0123} = +1$ for 4-dimensional.

History of the Motion of Extended Bodies

The present motivation to study relativistic motion of extend bodies is referred in the latter part of the introduction. However, motion of object on curved spacetimes have been considered from the old time. According to an article by Infeld [6], the studies on equation of relativistic motion took place since the year of 1927. In the studies on relativistic motion, it is admitted that we can classify the studies into two schools broadly speaking.

School of Fock and Papapetrou

The first school[1] is led by Fock and Papapetrou. Their idea is that (i) they first consider fluid on a spacetime, (ii) suppose that objects consist of the fluid, (iii) define the center of each object and (iv) derive equations of motion for the center. Fock and his collaborator did it by using post-Newtonian approximation, and Papapetrou did it in fully relativistic manner. Particularly, Papapetrou derived an equation, which is often called, referring to his name, as

[1] The term "school" is used in Fock's article. The author follows it in this chapter.

Papapetrou equation. His derivation can be considered as an natural extension of the procedure developed in the case of Newtonian gravity. We will start by reviewing it concretely.

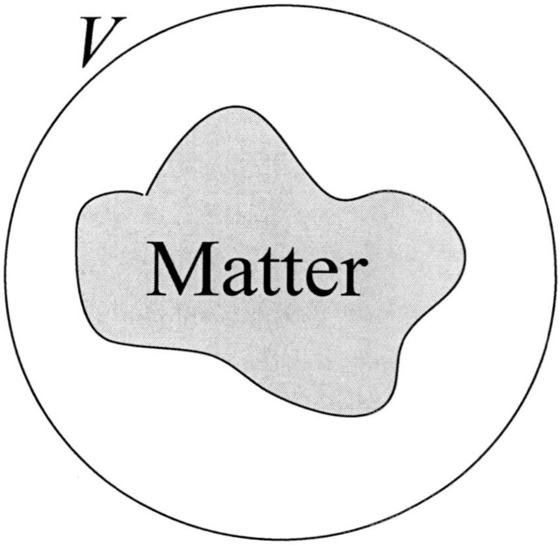

Figure 1. Region of integral. Integration is carried out over the inside of V.

First, we examine it in Newtonian gravity. We start from the following equation of motion for continuum in the Euler picture:

$$\frac{\partial}{\partial t}\left(\rho v^i\right) = -\frac{\partial P^{ij}}{\partial x^j} + \rho K^i, \tag{1}$$

where ρ is the density of the continuum matter, v^i is the velocity field, P^{ij} is the momentum flux density and K^i express the external gravitational field. Supposing that matter exists isolately, we can integrate this equation over a region V (fig.1) that enclose the matter distribution,

$$\frac{d}{dt}\int_V dV \rho \vec{v} = \int_V dV \rho \vec{K}. \tag{2}$$

Note that the term with P^{ij} becomes surface integral and has no influence on the above equation because the pressure and the velocity vanish at the boundary of V by the assumption. Here, we expand the external gravitational field K^i regarding an appropriate point as a center of the object,

$$\vec{K}(\vec{x}) = \vec{K}(\vec{X}) + \vec{K}_{,a}(\vec{X})y_a + \frac{1}{2!}\vec{K}_{,ab}(\vec{X})y_a y_b + \cdots, \tag{3}$$

where $\vec{X}$ means the center of the object and we set $\vec{x} = \vec{X} + \vec{y}$. The symbol "$,a$" in suffix means the derivative by x^a, which is often used in relativity. By substituting this equation into the right hand side of equation (2), we obtain,

$$\int_V dV \rho \vec{K} = \left(\int_V dV \rho\right)\vec{K}(\vec{X}) + \left(\int_V dV \rho y^a\right)\vec{K}_{,a}(\vec{X}) + \frac{1}{2}\left(\int_V dV \rho y^a y^b\right)\vec{K}_{,ab}(\vec{X}) + \cdots. \quad (4)$$

As usually, we choose the position $\vec{X}$ to the center-of-mass of the matter distribution,

$$\int_V dV \rho y^a = 0 \text{ or } \int_V dV \rho \vec{x} = M\vec{X}, \quad (5)$$

where we introduce $M \equiv \int_V dV \rho$ as the total mass of the object. Differentiating (5) by the time t and using the equation of continuity,

$$\frac{\partial \rho}{\partial t} + \frac{\partial(\rho v^i)}{\partial x^i} = 0, \quad (6)$$

the following equations are obtained,

$$M\dot{\vec{X}} = \int_V dV \frac{\partial \rho}{\partial t}\vec{x} = -\int_V dV \frac{\partial(\rho v^k)}{\partial x^k}\vec{x} = \int_V dV \rho \vec{v}. \quad (7)$$

Consequently, the equation of motion (2) becomes

$$M\ddot{\vec{X}} = M\vec{K}(\vec{X}) + \frac{1}{2}Q^{ab}\vec{K}_{,ab}(\vec{X}) + \cdots. \quad (8)$$

Thus, we can derive Newtonian equations of motion of objects with arbitrary multipole moments. Fock and his collaborators did the same calculation in post-Newtonian gravity [7], and Papapetrou did it in fully relativistic manner.

Next, we will review Papapetrou's work [8] that relates to our interest in this chapter. Papapetrou started from the relativistic versions of the Euler equations and of the equation of continuity,

$$\mathfrak{T}^{\mu\nu}{}_{;\nu} = 0, \quad (9)$$

where $T^{\mu\nu}$ is the energy momentum tensor *density* of matter, which relates to the energy momentum tensor $T^{\mu\nu}$ appearing in Einstein equation by the following equation,

$$\mathfrak{T}^{\mu\nu} = \sqrt{|g|}T^{\mu\nu}. \quad (10)$$

Equation (9) of $\mu = 0$ corresponds to the equation of continuity in Newtonian theory and $\mu = 1, 2, 3$ correspond to the (Newtonian) Euler equations. Writing equation (9) more concretely, it becomes

$$\mathfrak{T}^{\mu\nu}{}_{,\nu} + \Gamma^{\mu}_{\alpha\beta}\mathfrak{T}^{\alpha\beta} = 0. \tag{11}$$

We also use the following relation

$$\frac{\partial}{\partial x^{\rho}}\left(x^{\mu}\mathfrak{T}^{\nu\rho}\right) = \mathfrak{T}^{\nu\mu} - x^{\mu}\Gamma^{\nu}{}_{\rho\sigma}\mathfrak{T}^{\rho\sigma}, \tag{12}$$

which is derived from (11) in the below argument. Suppose that the matter distribution is local (fig.2). Note that the term By integrating these equations on a surface of simultaneity, we obtain

$$\frac{d}{dt}\int_{\Sigma} dV\mathfrak{T}^{\mu 0} + \int_{\Sigma} dV\Gamma^{\mu}_{\alpha\beta}\mathfrak{T}^{\alpha\beta} = 0, \tag{13}$$

$$\frac{d}{dt}\int_{\Sigma} dVx^{\mu}\mathfrak{T}^{\nu 0} = \int_{\Sigma} dV\mathfrak{T}^{\mu\nu} - \int_{\Sigma} dVx^{\mu}\Gamma^{\nu}{}_{\rho\sigma}\mathfrak{T}^{\rho\sigma}. \tag{14}$$

As well as Newtonian case, we expand the external gravitational field $\Gamma^{\mu}_{\alpha\beta}$ at the "mean" position X^{μ} of the matter distribution as follows,

$$\Gamma^{\mu}_{\alpha\beta}(x) = \Gamma^{\mu}_{\alpha\beta}(X) + \Gamma^{\mu}_{\alpha\beta,\rho}(X)y^{\rho} + \Gamma^{\mu}_{\alpha\beta,\rho\sigma}(X)y^{\rho}y^{\sigma} + \cdots. \tag{15}$$

In this equation, we set $x^{\mu} = X^{\mu} + y^{\mu}$. Now, suppose that the distribution of matter satisfies the following conditions

$$\int_{\Sigma} dV\, \mathfrak{T}^{\mu\nu} \neq 0, \tag{16}$$

$$\int_{\Sigma} dV\, y^{\rho}\mathfrak{T}^{\mu\nu} = 0. \tag{17}$$

The object satisfying these conditions are called as *single-pole particle* in Papapetrou's paper. Then, by taking only the first term of the right hand side of (15), we obtain the following geodesic equation after some calculations,

$$\frac{du^{\mu}}{d\tau} + \Gamma^{\mu}_{\alpha\beta}u^{\alpha}u^{\beta} = 0, \tag{18}$$

as equation of motion of the object.

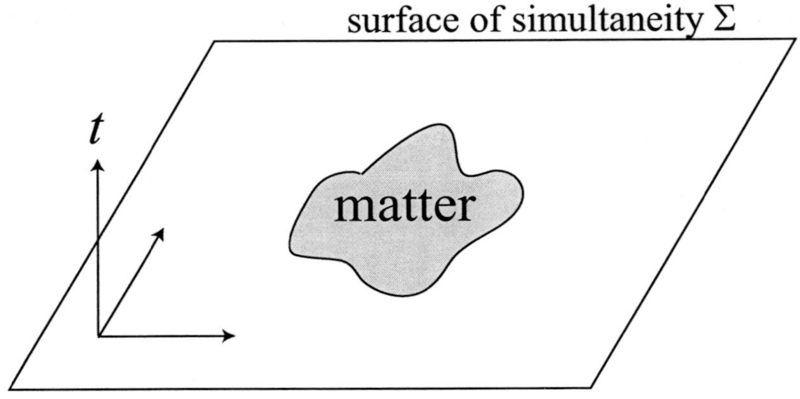

Figure 2. Image of distribution of matter and surface of simultaneity

If we take the first and second terms in the right hand side of (15) and suppose the following conditions on distribution of matter,

$$\int_{\Sigma} dV\ \mathfrak{T}^{\mu\nu} \neq 0, \tag{19}$$

$$\int_{\Sigma} dV\ y^{\rho}\mathfrak{T}^{\mu\nu} \neq 0, \tag{20}$$

$$\int_{\Sigma} dV\ y^{\rho}y^{\sigma}\mathfrak{T}^{\mu\nu} = 0, \tag{21}$$

we can obtain below after *using some tricks*,

$$\frac{D}{D\tau}\left(m^{\mu} + \frac{DS^{\mu\nu}}{D\tau}u_{\nu} \right) + \frac{1}{2}S^{\alpha\beta}u^{\sigma}R^{\mu}{}_{\beta\sigma\alpha} = 0, \tag{22}$$

as the equation of motion. This equation is often referred as Papapetrou equation. Here $\frac{D}{D\tau}$ means covariant time-derivative along the world line of the particle and $S^{\mu\nu}$ is the spin angular momentum of the particle.

There are some weak points in this Papapetrou's procedure. They are caused by the integration over a surface of simultaneity. First, usage of the integration over a surface of simultaneity makes the procedure not to be covariant, though the final result is covariant. Therefore, this derivation seems to be imperfect and complicated. This is the meaning that we wrote "*using some tricks*" just above. It is really this point. Indeed, Papapetrou consumed much of his paper to this. He discussed coordinate transformation to a covariant form in detail. The second problem in integrating over a simultaneous surface is that we can not mathematically justify the integration of tensor over a non-local region. This is related to the first matter and it is widely admitted that there is no sense in direct comparison of tensor components at any different points. If we want it to make sense, the argument is limited to a special coordinates where the integration is operated. However, these issues were solved by

Dixon [9], [10], [11]. Dixon developed a covariant procedure by using bi-tensor in integration of tensor over a region. The third point is that a surface of simultaneity is a set of points that are related non-causally to each other. The physical quantities, energy momentum tensor, on these point are integrated and the integrated result determines the motion of the objects. From the view of causality, this is disagreeable. Indeed, the non-causality seems to vanish apparently in the final form of equation of motion, but it is merely appearance. It is still true that non-causal information is included implicitly in equation of motion. Then, is there any way to avoid non-causality? This question is one of the motivations in starting our study.

School of Einstein, Infeld and Hoffmann

Next, let us review the ideas of Einstein and his collaborates. Their theory is known as Einstein-Infeld-Hoffmann formalism at present. They originally purposed to discuss relativistic many-body problem. Their theory about motion of objects was established in their article in 1938 [12]. They describe objects as singularities of gravitational field. By analyzing Einstein equation with post-Newtonian approximation, they derive equations of motion for the singularities. The singularities are used here based on the idea that when the external gravitational field outside the objects is same, the motion will be also same regardless of the composition and the structure of the objects. This idea will be also the one of multipole particle concept in this chapter. Their work was expanded to higher post-Newtonian order by Kimura, Ohta and their collaborates in 1970's and 1980's [13]. In this section, we will review the approach of this school according to Kimura et al.

The starting point of the approach of this school is just Einstein equation itself,

$$G^{\mu\nu} = \frac{8\pi G}{c^4} T^{\mu\nu}. \tag{23}$$

Here, we use the energy momentum tensor for point particles,

$$T^{\mu\nu} = -\sum_a \frac{m_a}{\sqrt{|g|}} \frac{dx_a^\mu}{dt} \frac{dx_a^\nu}{dt} \frac{dt}{d\tau_a} \delta^3(\vec{x} - \vec{z}_a(t)). \tag{24}$$

In this expression, we suppose a Cartesian type coordinates implicitly, $\vec{z}_a(t)$ means the position of the particle at the time t, and the suffix i distinguishes individual particles.

Let us solve Einstein equation with the energy momentum tensor by using post-Newtonian approximation. We assume that the motion is sufficiently slower than the speed of light and the gravitation is sufficiently weak. Under this assumption, we set

$$g_{\mu\nu} = \eta_{\mu\nu} + c^{-2} h_{\mu\nu}^{(2)} + c^{-3} h_{\mu\nu}^{(3)} + \cdots, \tag{25}$$

and we expand Einstein equation by inverse power of c. In the equation (25), the suffix in parenthesis, shows the order of inverse power of c: for example, (2) of $h_{\mu\nu}^{(2)}$ means at the order of c^{-2}.

As a consequence of post-Newtonian expansion, we get below at order c^{-2},

$$\Delta h_{00}^{(2)} = -8\pi \frac{G}{c^2}\sum_a m_a \delta^3(\vec{x}-\vec{z}_a), \tag{26}$$

$$\Lambda_{ij} h_{0j}^{(2)} = 0, \tag{27}$$

$$\Lambda_{ik:jl} h_{kl}^{(2)} = h_{00,ij}^{(2)} - 8\pi \frac{G}{c^2}\sum_a m_a \delta^3(\vec{x}-\vec{z}_a)\delta_{ij}, \tag{28}$$

where Λ_{ij}, $\Lambda_{ik:jl}$ are differential operators defined by,

$$\Lambda_{ij} \equiv \delta_{ij}\Delta - \partial_i \partial_j, \tag{29}$$

$$\Lambda_{ik:jl} \equiv \delta_{ik}\delta_{jl}\Delta + \delta_{kl}\partial_i\partial_j - \delta_{jl}\partial_i\partial_k - \delta_{ik}\partial_j\partial_l. \tag{30}$$

Solving equation (25), we obtain

$$h_{00}^{(2)} = \frac{2G}{c^2}\sum_a \frac{m_a}{r_a}, \tag{31}$$

$$h_{0i}^{(2)} = \partial I^{(2)}, \tag{32}$$

$$h_{ij}^{(2)} = \delta_{ij}\frac{2G}{c^2}\sum_a \frac{m_a}{r_a} + \partial_i J_j^{(2)} + \partial_j J_i^{(2)}, \tag{33}$$

where $I^{(2)}$ and $J_i^{(2)}$ are determined by a coordinate condition. Similarly, after troublesome calculations, we can derive equations for the metric at higher order c^{-3}, c^{-4}, $\cdots$, and solve the fields.

The important point of this procedure is that the integrable condition to the equation for $h_{kl}^{(4)}$ at the order c^{-4} gives the equation of motion in Newtonian gravity. This corresponds to that we can derive relativistic equation of motion from Bianchi identity in case of Einstein gravity. This fact indicates that when we consider the gravitational field of objects, the motion of the objects, which generates the gravitational field, is restricted by the condition that non-linear components of the field must smoothly join each other. This school described relativistic motion of many objects system in the way that we have seen in this section.

Present

How do we currently recognize relativistic motion of objects? As we referred in the introduction, some laser interferometric detectors for gravitational waves are under construction in some areas in the world. Their main targets are binary systems of compact object such as black holes, neutron stars and so on. The most frequently used methods in calculating the wave forms of the gravitational waves are the one using the post-Newtonian approximation. Although there are also elegant methods using black hole perturbation [14], they use a test particle approximation and they are not sufficient to apply to the targets that consist of objects with almost same masses. Therefore, people often use the post-Newtonian approximation methods [15]. In these methods, only geometric quantities are expanded by using post-Newtonian (+post-Minkowskian) approximation and quantities of matter are left original, i.e. fully relativistic. Finally, by substituting the energy momentum tensor to the post-Newtonian field equation, the fields are obtained and the equation of motion are derived.

In the argument of these post-Newtonian methods, people often use simple point particles (i.e. they are specified by only their mass.) The use of point particles simplifies the calculations significantly because the energy momentum tensor for the particles can be expressed by delta functions. This is an important merit in the extremely complicated post-Newtonian calculations. Furthermore, some works discuss the case that the object may have angular momentum by using the energy momentum tensor for spinning particles [16], [17], [18]. It may be possible to analyze post-Newtonian gravitational fields and motions fluid-mechanically supposing some equation of states, but it seems to be difficult to argue the dynamics without any assumption of high symmetries on the matter distribution.

Thus, we ask if we can generalize the description with point particles in order to describe the motion of various object with an arbitrary internal structure and also to simplify the post-Newtonian calculations. This question is the primary motivation of our study developed in this chapter.

IDEA OF MULTIPOLE PARTICLE

In this section, we bring up a basic idea of multipole particle. Probably, it is the easiest way to discuss it in electromagnetism. Therefore, let us first consider a static electric field.

The most fundamental law of electromagnetism is Coulomb's law for a point charge. Suppose that a point-like charge q is at a position $\vec{x} = \vec{x}_0$. Then, the static electric potential ϕ is derived by the following Coulomb's law

$$\nabla^2 \phi = -\frac{q}{4\pi\varepsilon_0} \delta^3(\vec{x} - \vec{x}_0). \tag{34}$$

This electric potential for a point charge is just a Green function of the equation (34). Because this equation is linear, we can superpose the solutions for the equation. Generally, when charges distributed such as the density $\rho(x)$, it is well known that the potential is given by

$$\nabla^2 \phi = -\frac{\rho(x)}{4\pi\varepsilon_0}. \tag{35}$$

Now, let us consider a situation that two point charges $\pm q$ are at positions $\vec{x} = \pm\vec{a}$. The charge density in this case is given by

$$\rho(x) = q\delta^3(\vec{x} - \vec{a}) - q\delta^3(\vec{x} + \vec{a}). \tag{36}$$

Let us make the two charge be closer to each other $\vec{a} \to \vec{0}$, keeping $\vec{p} \equiv q\vec{a}$ constant. Although there are Dirac's delta function in the charge density, we obtain the following density by a formal Taylor expansion,

$$\rho(x) = -2(\vec{p} \cdot \nabla)\delta^3(x) + O(q\,|\,\vec{a}\,|^2), \tag{37}$$

where $\vec{p}$ is an electric dipole vector. The equation for the point source (37), expressed by a delta function with a differential operator along $\vec{p}$, is the equation of (34) with differentiation along $\vec{p}$,

$$\nabla^2 \left\{ \frac{2(\vec{p} \cdot \nabla)}{q} \right\} \phi = -2(\vec{p} \cdot \nabla)\delta^3(x). \tag{38}$$

Therefore, if we know a solution ϕ_0 to equation (34) with $q = 1$ under an adequate boundary condition, we can obtain an electric potential ϕ_1 for the point source (37) as follows,

$$\phi_1(x) = -2(\vec{p} \cdot \nabla)\phi_0(x). \tag{39}$$

This idea can be applied to higher electric multipoles in the same way.

In order to avoid complicated expression, we do not use the notation such as $(\vec{p} \cdot \nabla)$ in case of higher multipoles. We employ tensors to describe higher multipoles as follows,

$$\begin{aligned} Q &\equiv q, \\ Q^i &\equiv p^i, \\ Q^{ij} &\equiv r^i_{(1)} r^j_{(2)}, \\ &\cdots. \end{aligned}$$

Using these multipole tensors, we can express a point source with higher multipole distribution of charge as follows,

$$\rho(x) = Q\delta^3(\vec{x}) + Q^i \partial_i \delta^3(\vec{x}) + Q^{ij} \partial_i \partial_j \delta^3(\vec{x}) \cdots \equiv \sum_{l=0}^{\infty} Q^L \partial_L \delta^3(\vec{x}), \tag{40}$$

where we introduce the multi-index notation and use so-called Einstein's summation rule. The meaning of the multi-index notation is that the index L in equation (40) shows a set of l indices such as

$$L = i_1 i_2 \cdots i_l .$$

In other words, the following equation

$$Q^L \partial_L = \sum_{i_1} \cdots \sum_{i_l} Q^{i_1 \cdots i_l} \partial_{i_1} \cdots \partial_{i_l}, \tag{41}$$

concretely shows the meaning of the multi-index notation. Of course, we can choose any other character than i 's because it is a mere dummy index. Using this notation, we can write an electric potential for the source (40) as follows,

$$\phi = \left(\sum_{l=0}^{\infty} Q^L \partial_L \right) \phi_0 . \tag{42}$$

Strictly speaking, the Q^L 's themselves, defined above, are not exact multipole moments though they relate closely to electric multipole moments. It is because we do not consider any symmetries among the indices.

First, we can generally suppose that Q^L is a perfectly symmetric tensor because Q^L always appears with ∂_L and the differential operators ∂_i are commutable with each other:

$$Q^{ab \cdots c} = Q^{(ab \cdots c)} . \tag{43}$$

Next, notice that equation (34) is of rotational symmetry on account of an isotropic property of Coulomb force. Consequently, we expect that any solutions of (34) can be expanded by representation of SO(3). That is, Q^L must be an element of the representation space of SO(3). Then, the irreducibility of the representation assures that Q^L 's are multipole moments. Remember that the concept of the multipole moments is a classification by the rotational group SO(3). For example, $Q^L = Q$ of $l = 0$ is an irreducible representation of angular momentum quantum number $j = 0$ and $Q^L = Q$ of $l = 1$ is that of $j = 1$. However, Q^{ij} of $l = 2$ is not an irreducible representation of $j = 2$, because the trace part of Q^{ij} behaves as a representation of $j = 0$. Then, we must remove this component from Q^{ij} in order to make $j = 2$ representation. Namely, we set

$$Q^{ij} = \frac{1}{3} Q^{ab} \delta_{ab} \delta^{ij} + \left(Q^{ij} - \frac{1}{3} Q^{ab} \delta_{ab} \delta^{ij} \right), \tag{44}$$

and obtain the irreducible representation of $j = 2$ in the parenthesis in this equation. This is exactly the well-known definition of the quadrupole moment. If we make $j = 0$ representation absorb all $j = 0$ part of higher rank representations from the start, Q^L has a property that any trace of Q^L will vanish. We call the property of Q^L, perfectly symmetric and trace free, as symmetric-trace-free (STF). To sum up, the three dimensional STF tensors make an irreducible representation space of SO(3). This is a mathematical definition of multipole moments.

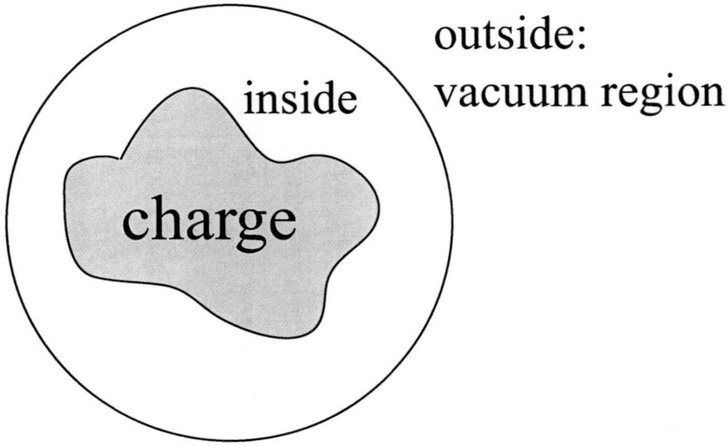

Figure 3. Distribution of charge and distinction of regions

Let us set our story back to the electric field. We suppose that the charge distribution is local and it is a vacuum distant enough from the charge (fig.3). Note that the source term, we also adopt the boundary condition that electric potential, vanishes at the infinity. Under these assumptions, the equation for the electric potential becomes the Laplace equation,

$$\nabla^2 \phi = 0, \tag{45}$$

in the vacuum region (fig,3). Under the given boundary condition, a spherical symmetric solution is

$$\phi_0 = \frac{Q}{r}, \tag{46}$$

and other solutions can be obtained by operating adequate differential operators to this spherical symmetric solution ϕ_0. The general solution to the Laplace equation is a linear combination of these solutions. That is, in the vacuum region sufficiently distant from the distribution of charge (fig.3), the general solution is expressed by

$$\phi = \sum_{l=0}^{\infty} Q^L \partial_L \left(\frac{1}{r} \right), \tag{47}$$

where Q^L has STF symmetry as referred above. Once we have this general solution, by using the following relation

$$\nabla^2 \left(\frac{1}{r} \right) = -4\pi \delta^3 (\vec{x}), \tag{48}$$

we can obtain the point-like source. Therefore, we can conclude the following theorem:

Theorem 1. There is a point charge that can create, in a vacuum region, the same electric field which is created by arbitrary local distribution of electric charge.

So far, we have only examined a static electric field. We will hereafter consider a case that electric multipole moments have time dependencies. The equation for time dependent fields is given by a Klein-Gordon equation,

$$\Box\phi = -\frac{\rho}{\varepsilon_0}. \tag{49}$$

Since a retarded Green function of Klein-Gordon equation (49) is given by

$$G(\vec{r},t) = \frac{1}{4\pi r}\delta(t-r), \tag{50}$$

we change the potential (47) to

$$\phi(t,\vec{r}) = \sum_{l=0}^{\infty}\partial_L\left[\frac{Q^L(t-r)}{r}\right]. \tag{51}$$

(Recall a Green function of Laplace equation is $\frac{1}{r}$.) Thereupon, we have

$$\Box\phi = -\frac{1}{\varepsilon_0}\sum_{l=0}^{\infty}\partial_L\left[Q^L(t)\delta^3(\vec{r})\right], \tag{52}$$

and obtain an expression for the point charge with arbitrarily time-dependent multipole moments.

What is the advantage of replacing an extended distribution with a point charge? The advantage we consider is that there is no necessity of integrating the charge distribution in order to define multipole moments. In the point charge description, we do not need any detail information of charge distribution since we define the multipole moments by the field at a distance. To be exact, this method defines the multipole moments of the *electric field* but not the multipole moments of the *distribution of electric charge*. However, as seen above, these two moments are equivalent to each other in case of electromagnetism. Furthermore, there is a point which requires some attention. If we suppose distribution of electric charge, we can uniquely obtain the electric field under a certain boundary condition. However, the inverse argument does not hold. Generally, there are many charge densities that give the same electric field. Namely, even when we have the multipole moments of the electric field, we can not determine the charge distribution. However, this is not a defect because we do not need to know such charge distribution as far as we consider the motion of the charge. All physical phenomenon has to occur only through the electric fields around the charge.

GRAVITATIONAL FIELD IN SPECIAL RELATIVITY

In this section, we apply the idea of the particle-like source in the last section to gravitational fields. Since the relativistic gravity is highly non-linear system and very complicated, we first consider a linearized gravitation.

The basic equation of general relativity is usually written in an elegant form in term of differential geometry, so-called Einstein equation:

$$G^{\mu\nu} = \frac{8\pi G}{c^4} T^{\mu\nu}. \tag{53}$$

However, there is another type of expression for Einstein equation. It is a Klein-Gordon type form of Einstein equation,

$$\Box h^{\mu\nu} = -\frac{16\pi G}{c^4}(-g)\left(T^{\mu\nu} + t_{\mathrm{LL}}^{\mu\nu}\right) + \left(h^{\mu\nu}{}_{,\rho\sigma} h^{\rho\sigma} - h^{\mu\rho}{}_{,\sigma} h^{\nu\sigma}{}_{,\rho}\right), \tag{54}$$

where $\Box$ is d'Alembertian by Lorenz metric $\eta^{\mu\nu} = \mathrm{diag}(-1,1,1,1)$ of Minkowski spacetime, $h^{\mu\nu}$ is defined from the metric $g^{\mu\nu}$ of the curved spacetime as follows,

$$h^{\mu\nu} = \eta^{\mu\nu} - \sqrt{|g|} g^{\mu\nu},$$

and $t_{\mathrm{LL}}^{\mu\nu}$ is Landau-Lifshitz pseudo-tensor. Furthermore, we impose de Donder gauge condition

$$h^{\mu\nu}{}_{,\nu} = 0, \tag{55}$$

in order to fix the freedom of coordinate transformation. Additionally, we raise and lower indices by Lorenz metric. We emphasize that this Klein-Gordon type equation is perfectly equivalent to the original Einstein equation (53).

Suppose that the energy momentum density of matter is sufficiently small, that is, the gravitational field is weak enough and the motion of object is sufficiently slower than the speed of light. Under this assumption, we can drop out the non-linear terms of $h^{\mu\nu}$ from equation (54). After this approximation, we obtain the linearized Einstein equation that describes weak gravitational fields as below,

$$\Box h^{\mu\nu} = -\frac{16\pi G}{c^4} T^{\mu\nu}. \tag{56}$$

This equation is of the same structure as the equation for the electric potential considered in the last section except that there are tensor indices with h and T. We also suppose that distribution of matter is local (fig.4). Therefore, general solutions to (56) in a vacuum region are given by

$$h^{\mu\nu} = \sum_{l=0}^{\infty} \Theta^{(\mu\nu)L}(t)\partial_L\left(\frac{1}{r}\right). \tag{57}$$

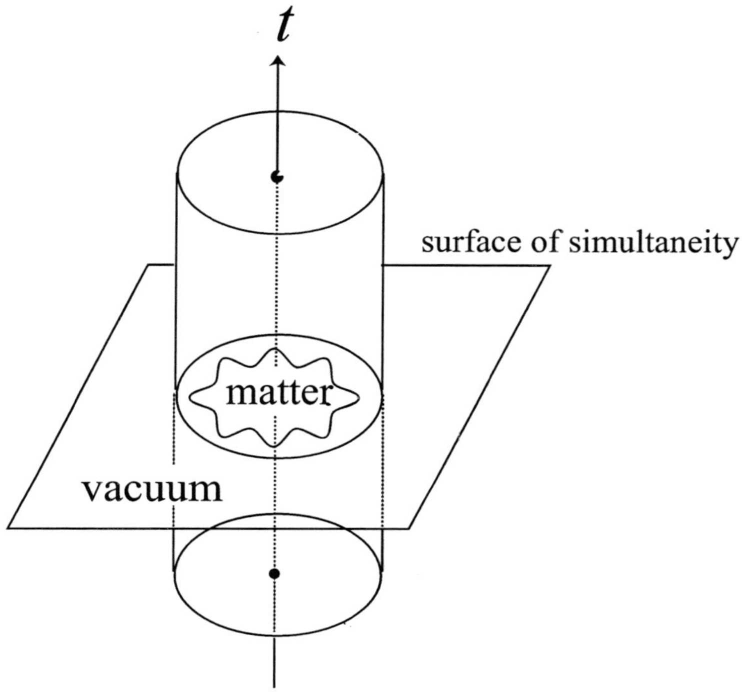

Figure 4. Distribution of matter. It is vacuum outside the tube and matter exists only inside of the tube.

Note that there is an essential difference between the electric potential ϕ and this gravitational field $h^{\mu\nu}$: the electric potential is a scalar, but the gravitational field $h^{\mu\nu}$ is itself a tensor. For instance, when the rotational group SO(3) acts on $h^{\mu\nu}$, h^{00} behaves as an irreducible representation of $j=0$, h^{0i} does a $j=1$ irreducible representation and h^{ij} behaves as a mixture of $j=0$ and $j=2$ representations. Because of this fact, from the view point of representation of SO(3), $\Theta^{(00)L}$ is a direct product of $j=0$ and $j=l$ representations, $\Theta^{(0i)L}$ is a direct product of $j=1$ and $j=l$ representations, and $\Theta^{(ij)L}$ is a direct product of $j=0,2$ and $j=l$ representations. Except the first one, these Θ 's are not irreducible representations of SO(3) in general. Therefore, we can not immediately interpret Θ 's as multipole moments. In order to define multipole moments, we must decompose $\Theta^{(\mu\nu)L}$ into irreducible representations of SO(3).

The way to decompose $\Theta^{(\mu\nu)L}$ into irreducible representation is exactly same as the way of composing angular momentum in quantum mechanics. Since $\Theta^{(00)L}$ is a direct product of $j=0$ representation (1) and $j=l$ representation $(2l+1)$, this $\Theta^{(00)L}$ itself is a $j=l$ irreducible representation:

$$1 \otimes (2l+1) = (2l+1). \tag{58}$$

$\Theta^{(0i)L}$, which is a direct product of $j=1$ (3) and $j=l$ $(2l+1)$, is decomposed into three irreducible representation $j=l-1$ $(2l-1)$, $j=l$ $(2l+1)$ and $j=l+1$ $(2l+3)$:

$$3 \otimes (2l+1) = (2l-1) \oplus (2l+1) \oplus (2l+3). \tag{59}$$

$\Theta^{(ij)L}$, where the part of (ij) is mixture of $j=0$ and $j=2$ representation, is decomposed into six irreducible representation of $j=l-2$ $(2l-3)$, $j=l-1$ $(2l-1)$, two $j=l$'s $(2l+1)$, $j=l+1$ $(2l+3)$ and $j=l+2$ $(2l+5)$:

$$\begin{aligned}\{1 \oplus 5\} \otimes (2l+1) = (2l+1) \oplus \{&(2l-3) \oplus (2l-1) \\ &\oplus (2l+1) \oplus (2l+3) \oplus (2l+5)\}.\end{aligned} \tag{60}$$

In order to construct a concrete expression for multipole moments, we combine STF tensors and two SO(3)-invariant tensors, δ^{ij} and ε^{ijk}, as given below [19],

$$h^{00} = \frac{4G}{c^4}\sum_{l=0}^{\infty}\partial_L\left[\frac{M^L(t_{\text{ret}})}{r}\right], \tag{61}$$

$$\begin{aligned} h^{0i} &= \frac{4G}{c^4}\sum_{l=1}^{\infty}\left\{\partial_{L-1}\left[\frac{P^{iL-1}(t_{\text{ret}})}{r}\right] + \partial_{aL-1}\left[\frac{\varepsilon^{iam}S^{mL-1}(t_{\text{ret}})}{r}\right]\right\} \\ &\quad + \frac{4G}{c^4}\sum_{l=0}^{\infty}\partial_{aL}\left[\frac{\delta^{ia}Q^L(t_{\text{ret}})}{r}\right], \end{aligned} \tag{62}$$

$$\begin{aligned} h^{ij} &= \frac{4G}{c^4}\sum_{l=0}^{\infty}\delta^{ij}\partial_L\left[\frac{A^L(t_{\text{ret}})}{r}\right] \\ &\quad + \frac{4G}{c^4}\sum_{l=2}^{\infty}\left\{\partial_{L-2}\left[\frac{B^{ijL-2}(t_{\text{ret}})}{r}\right] + \partial_{aL-2}\left[\frac{\varepsilon^{am(i}C^{j)mL-2}(t_{\text{ret}})}{r}\right]\right\} \\ &\quad + \frac{4G}{c^4}\sum_{l=1}^{\infty}\left\{\partial_{aL-1}\left[\frac{\delta^{a(i}D^{j)L-1}(t_{\text{ret}})}{r}\right] + \partial_{abL-1}\left[\frac{\delta^{a(i}\varepsilon^{j)bm}E^{mL-1}(t_{\text{ret}})}{r}\right]\right\} \\ &\quad + \frac{4G}{c^4}\sum_{l=0}^{\infty}\partial_{abL}\left[\frac{\delta^{ia}\delta^{jb}F^L(t_{\text{ret}})}{r}\right], \end{aligned} \tag{63}$$

where $t_{\text{ret}} \equiv t - r$. These express the most general gravitational field, but it has some degrees of freedom that results from the covariant properties of coordinate transformation, which general relativity originally has. This freedom is often referred as *gauge freedom*.

This freedom can be fixed as follows. First, according to de Donder condition (55) imposed in the derivation of linearized Einstein equation, the following relations among the tensors in (61)-(63) must hold,

$$\dot{M}^L + P^L + \ddot{Q}^L = 0, \tag{64}$$

$$\dot{P}^L + B^L + \frac{1}{2}\ddot{D}^L = 0, \tag{65}$$

$$\dot{S}^L + \frac{1}{2}C^L + \frac{1}{2}\ddot{E}^L = 0, \tag{66}$$

$$\dot{Q}^L + A^L + \frac{1}{2}D^L + \ddot{F}^L = 0, \tag{67}$$

where the dots above the tensors mean derivative by the time. However, de Donder gauge condition does not completely fix the gauge freedom as well as the harmonic gauge in electromagnetism, and it remains some gauge freedom. Then, let us consider the following gauge transformation: we change the coordinate $x^{\mu}_{(\text{old})}$ to a new coordinate $x^{\mu}_{(\text{new})}$ by a infinitesimal distance ξ^{μ} as below,

$$x^{\mu}_{(\text{old})} \to x^{\mu}_{(\text{new})} = x^{\mu}_{(\text{old})} + \xi^{\mu}. \tag{68}$$

Under this transformation, the linearized gravitational field is transformed by

$$h^{\mu\nu}_{(\text{old})} \to h^{\mu\nu}_{(\text{new})} = h^{\mu\nu}_{(\text{old})} + \xi^{\mu,\nu} + \xi^{\nu,\mu} - \eta^{\mu\nu}\xi^{\rho}{}_{,\rho}. \tag{69}$$

Let us adopt the following functions as gauge function ξ^{μ},

$$\xi^0 = \frac{4G}{c^4}\sum_{l=0}^{\infty}\left\{\partial_L\left[\frac{Q^L}{r}\right] + \frac{1}{2}\partial_L\left[\frac{\dot{F}}{r}\right]\right\}, \tag{70}$$

$$\xi^i = \frac{4G}{c^4}\sum_{l=1}^{\infty}\left\{\frac{1}{2}\partial_{L-1}\left[\frac{D^{iL-1}}{r}\right] + \partial_{aL-1}\left[\frac{\varepsilon^{iam}E^{mL-1}}{r}\right]\right\} + \frac{4G}{c^4}\sum_{l=0}^{\infty}\frac{1}{2}\partial_{aL}\left[\frac{\delta^{ia}F^L}{r}\right]. \tag{71}$$

After this transformation, the new linearized gravitational field has no terms that correspond to A^L, D^L, E^L and F^L. Consequently, we obtain the following gravitational field,

$$h^{00} = \frac{4G}{c^4}\sum_{l=0}^{\infty}\partial_L\left[\frac{M^L(t_{\rm ret})}{r}\right], \tag{72}$$

$$h^{0i} = \frac{4G}{c^4}\sum_{l=1}^{\infty}\left\{-\partial_{L-1}\left[\frac{\dot{M}^{iL-1}(t_{\rm ret})}{r}\right]+\partial_{aL-1}\left[\frac{J^{ia\cdot L-1}(t_{\rm ret})}{r}\right]\right\}, \tag{73}$$

$$h^{ij} = \frac{4G}{c^4}\sum_{l=2}^{\infty}\left\{\partial_{L-2}\left[\frac{\ddot{M}^{ijL-2}(t_{\rm ret})}{r}\right]+2\partial_{aL-2}\left[\frac{\dot{J}^{a(i\cdot j)L-2}(t_{\rm ret})}{r}\right]\right\}, \tag{74}$$

whose gauge freedom is perfectly fixed. In these equations, we introduce a new symbol

$$J^{ia\cdot L-1} \equiv \varepsilon^{iam}S^{mL-1}, \tag{75}$$

instead of S^L for notational simplicity. We refer the point $r = 0$ as a *center* of the gravitational field.

Now, we will name the multipole moments. As you see, M^L is in the (00)-component of $h^{\mu\nu}$. This (00)-component corresponds to mass density in Newtonian theory. Accordingly, we call M^L *mass multipole moments* or *mass multipoles*. On the other hand, there is no correspondence of S^L or $J^{ia\cdot L}$ in Newtonian theory. Because these moments is in the h^{0i} component with ε^{ijk}, these create a gravitational field that originates in the angular momentum of the object. Accordingly, we call them *spin multipole moments*. The multipole moments of the gravitational field in h^{0i} component are often referred by *current multipole moments*. However, we use *spin* multipole moments in order to emphasize that they are strongly related to the spin angular moment of the object.

Thus, we obtain the most general and perfectly gauge-fixed vacuum solution of linearized gravitational field. We can immediately make the energy momentum tensor for a particle with arbitrary multipole moments, but there remains a problem to be solved before getting energy momentum tensor. This problem concerns with the freedom of selection of a surface of simultaneity. It does not exist in Newtonian theory. Many of textbooks of relativity define the surface of simultaneity by propagation of light [20]. The coordinates defined by light propagation are rectangular coordinates where the temporal axis and the spatial axes are orthogonal to each other. However, this selection is one of the convenient coordinates and there is no physical reason why we must select it. By nature, the surface of simultaneity is a spatial supersurface and arbitrary two points have no causal relationship. Therefore, we can choose arbitrary spatial surface as a surface of simultaneity, but we customary and tacitly choose the rectangular coordinate. We will discuss this problem in the following section.

OBLIQUE FRAME

In this section, we examine the possibilities of introducing oblique coordinates. As we have indicated in the last section, there is arbitrariness choosing a surface of simultaneity for the center of the gravitational field. If we do not matter whether it is good or bad as a global coordinate, we can choose any spatial surface locally as the surface of simultaneity. There also exist arbitrariness choosing the degree of curved surface, but this does not matter, because we will consider a *point* particle and it is enough to discuss only the neighborhood of the particle. Namely, it is sufficient to consider planes only. Consequently, our geometrical question becomes which plane we should choose as a surface of simultaneity. Such spatial planes are specified by a timelike 4-vector and the simultaneity is determined by the geometrical relation between the 4-velocity of the particle and the 4-vector normal to the surface.

Then, let us consider two tetrads that are related to each other by a Lorenz transformation. The one of the tetrads is determined by the 4-velocity of the particle and the other is determined by the 4-vector normal to the *oblique* simultaneous surface. We set the 4-velocity of the particle as u^μ and three vectors normal to it as λ_i^μ, ($i = 1,2,3$). Here, we suppose that all λ_i are normal to each other. We call a tetrad $\{u, \lambda_i\}$ constructed by these four vectors as (u,λ) *-frame* hereafter. In the same way, we set the oblique vector, normal to the surface, as v^μ, and three vectors normal to v^μ as e_i^μ, ($i = 1,2,3$). These vectors e_i^μ, ($i = 1,2,3$) are also normal to each other. We call a tetrad $\{v, e_i\}$ constructed by these four vectors, as (e,λ) *-frame*. Let us write the Lorenz transformation between these two frames as follows,

$$x^\mu = \Lambda^\mu{}_\nu \bar{x}^\nu, \tag{76}$$

where x^μ is the coordinate in the (u,λ) -frame, and $\bar{x}^\mu$ is the coordinate in the (v,e) -frame. We can construct a *oblique* frame by combining these two frames. That is, we use u as the direction toward the future and e_i as the spatial basis. We call a tetrad $\{u, e_i\}$ constructed by these four vectors as (u,e) *-frame*. We show the relation among (u,λ) -frame, (v,e) -frame and (u,e) -frame in figure 5.

Our purpose here is to expand the gravitational field with multipole moments in this oblique (u,e) -frame in the same way in the last section. However, since the oblique frame complicates the gauge transformation, it is difficult to find out convenient gauge functions in this case. Therefore, we introduce another way by which we can avoid considering any complicated gauge transformation. Our approach makes the argument easier, and furthermore, gauge fixing will be assured.

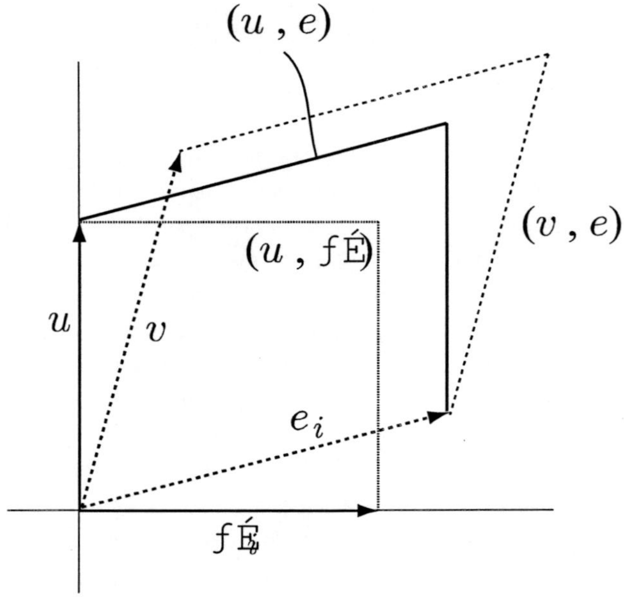

Figure 5. Definition of frames.

We start from the equation for the gravitational field in this oblique frame. The equation for a linearized gravitational field is given by

$$\tilde{\Box} h^{\mu\nu} = 0, \tag{77}$$

where $\tilde{\Box}$ is the d'Alembertian in (u, e) -frame. We distinguish the coordinates in this (u, e) -frame from those in other frame by putting a tilde $\tilde{}$. We need only vacuum solutions of (77) and we set the energy momentum tensor to zero. The solution for the linearized gravity in this (u, e) -frame is also given by the almost same form in (u, λ) -frame as follows,

$$\tilde{h}^{00} = \frac{4G}{c^4} \sum_{l=0}^{\infty} \tilde{\partial}_L \left[\frac{M^L(\tilde{t}_{\text{ret}})}{\tilde{r}} \right], \tag{78}$$

$$\begin{aligned} \tilde{h}^{0i} &= \frac{4G}{c^4} \sum_{l=1}^{\infty} \left\{ \tilde{\partial}_{L-1} \left[\frac{P^{iL-1}(\tilde{t}_{\text{ret}})}{\tilde{r}} \right] + \tilde{\partial}_{aL-1} \left[\frac{\varepsilon^{iam} S^{mL-1}(\tilde{t}_{\text{ret}})}{\tilde{r}} \right] \right\} \\ &+ \frac{4G}{c^4} \sum_{l=0}^{\infty} \tilde{\partial}_{aL} \left[\frac{\delta^{ia} Q^L(\tilde{t}_{\text{ret}})}{\tilde{r}} \right], \end{aligned} \tag{79}$$

$$\tilde{h}^{ij} = \frac{4G}{c^4}\sum_{l=0}^{\infty}\delta^{ij}\tilde{\partial}_L\left[\frac{A^L(\tilde{t}_{\text{ret}})}{\tilde{r}}\right]$$
$$+\frac{4G}{c^4}\sum_{l=2}^{\infty}\left\{\tilde{\partial}_{L-2}\left[\frac{B^{ijL-2}(\tilde{t}_{\text{ret}})}{\tilde{r}}\right]+\tilde{\partial}_{aL-2}\left[\frac{\varepsilon^{am(i}C^{j)mL-2}(\tilde{t}_{\text{ret}})}{\tilde{r}}\right]\right\}$$
$$+\frac{4G}{c^4}\sum_{l=1}^{\infty}\left\{\tilde{\partial}_{aL-1}\left[\frac{\delta^{a(i}D^{j)L-1}(\tilde{t}_{\text{ret}})}{\tilde{r}}\right]+\tilde{\partial}_{abL-1}\left[\frac{\delta^{a(i}\varepsilon^{j)bm}E^{mL-1}(\tilde{t}_{\text{ret}})}{\tilde{r}}\right]\right\} \quad (80)$$
$$+\frac{4G}{c^4}\sum_{l=0}^{\infty}\tilde{\partial}_{abL}\left[\frac{\delta^{ia}\delta^{jb}F^L(\tilde{t}_{\text{ret}})}{\tilde{r}}\right].$$

Here, the valuables with tildes $\tilde{}$ are ones in (u,e)-frame. We pay attention to $\tilde{t}_{\text{ret}}$. Although $\tilde{t}_{\text{ret}} \neq \tilde{t}-\tilde{r}$ in general, we can concretely express $\tilde{t}_{\text{ret}}$ by $\tilde{t}$ and $\tilde{r}$ if we know the relation between u^μ and v^μ. However, we do not need any concrete expression of $\tilde{t}_{\text{ret}}$ in the following argument. What we must make sure in the general solution (77) is that this solution is exactly an irreducible representation of SO(3) and how we can fix gauge freedom of the gravitational field. These two problems are able to be solved simultaneously as we will see in below.

What we need to focus in solving those problems should be obvious. Namely, we notice that the solution (61)-(63) in (u,λ)-frame and the solution (77) in (u,e)-frame have exactly the same form. Accordingly, we consider a one-to-one mapping between these solutions as follows,

$$T:\begin{cases} h^{\mu\nu} \\ t_{\text{ret}}, r \\ \partial_\mu \end{cases} \rightarrow \begin{cases} \tilde{h}^{\mu\nu} \\ \tilde{t}_{\text{ret}}, \tilde{r}, \\ \tilde{\partial}_\mu \end{cases} \quad (81)$$

for example,

$$T^{-1}\left(\sum_{l=0}^{\infty}\tilde{\partial}_L\left[\frac{M^L(\tilde{t}_{\text{ret}})}{\tilde{r}}\right]\right)=\sum_{l=0}^{\infty}\partial_L\left[\frac{M^L(t_{\text{ret}})}{r}\right]. \quad (82)$$

It is clear that this mapping T makes the solution in (u,λ)-frame correspond to the solution in (u,e)-frame by one-to-one. Hence, we combine the mapping T and the representation ρ of SO(3) in (u,λ)-frame and make the following mapping $\tilde{\rho}$,

$$\tilde{\rho} \equiv T^{-1}\rho T. \quad (83)$$

Then, $\tilde{\rho}$ is a representation of SO(3) in (u,e)-frame. If we assure that the gauge freedom is fixed in one frame, we know that it is also true in the other frame because the property of one-to-one.

After all, by using the above mapping T, we can obtain the gauge-fixed linearized gravitational field in (u,e)-frame as follows,

$$\tilde{h}^{00}=\frac{4G}{c^4}\sum_{l=0}^{\infty}\tilde{\partial}_L\left[\frac{M^L(\tilde{t}_{\text{ret}})}{\tilde{r}}\right], \tag{84}$$

$$\tilde{h}^{0i}=\frac{4G}{c^4}\sum_{l=1}^{\infty}\left\{-\tilde{\partial}_{L-1}\left[\frac{\dot{M}^{iL-1}(\tilde{t}_{\text{ret}})}{\tilde{r}}\right]+\tilde{\partial}_{aL-1}\left[\frac{J^{ia\cdot L-1}(\tilde{t}_{\text{ret}})}{\tilde{r}}\right]\right\}, \tag{85}$$

$$\tilde{h}^{ij}=\frac{4G}{c^4}\sum_{l=2}^{\infty}\left\{\tilde{\partial}_{L-2}\left[\frac{\ddot{M}^{ijL-2}(\tilde{t}_{\text{ret}})}{\tilde{r}}\right]+2\tilde{\partial}_{aL-2}\left[\frac{\dot{J}^{a(i\cdot j)L-2}(\tilde{t}_{\text{ret}})}{\tilde{r}}\right]\right\}. \tag{86}$$

This is exactly an expression that corresponds with the solution in (u,λ)-frame by the mapping T.

Multipole Particle in Special Relativity

We obtain a set of the general solutions for the linearized gravitational field through the argument in the last section. Using these solutions, we can derive an energy momentum tensor for a point particle in case of the gravitational field as well as the electric field.

By substituting (84)-(86) for the following relation,

$$\tilde{\Box}\left(\frac{4G}{c^4}\frac{M(\tilde{t}_{\text{ret}})}{\tilde{r}}\right)=-\frac{16\pi G}{c^4}M(\tilde{t})\delta^3(\tilde{r}), \tag{87}$$

we get the energy momentum tensor for a point particle as follows,

$$\tilde{T}^{00}=\sum_{l=0}^{\infty}\partial_L\left[M^L(t)\delta^3(r)\right], \tag{88}$$

$$\tilde{T}^{0i}=\sum_{l=1}^{\infty}\left\{-\partial_{L-1}\left[\dot{M}^{iL-1}(t)\delta^3(r)\right]+\partial_{aL-1}\left[J^{ia\cdot L-1}(t)\delta^3(r)\right]\right\}, \tag{89}$$

$$\tilde{T}^{ij}=\sum_{l=2}^{\infty}\left\{\partial_{L-2}\left[\ddot{M}^{ijL-2}(t)\delta^3(r)\right]+2\partial_{aL-2}\left[\dot{J}^{a(i\cdot j)L-2}(t)\delta^3(r)\right]\right\}. \tag{90}$$

We call this particle *a multipole particle.*

Incidentally, what is the coordinate where this energy momentum tensor is described? It is obviously a frame where the particle rests at the origin, that is, the comoving frame of the particle. Our goal is to describe the motion of this particle, then we must change the coordinate to a general coordinate. The change of the coordinates is achieved by

$$T^{\mu\nu} = u^{\mu}u^{\nu}\tilde{T}^{00} + 2u^{(\mu}e_i^{\nu)}\tilde{T}^{0i} + e_i^{\mu}e_j^{\nu}\tilde{T}^{ij}. \tag{91}$$

When we perform this coordinate transformation, we must take notice of some points. First, it is a point concerning the Dirac's delta function. Because 3-dimensional delta function $\delta^3(\tilde{r})$ is not covariant, we must rewrite the delta functions in the energy momentum tensor by 4-dimensional covariant delta function as following:

$$\delta^3(\tilde{r}) = \int \mathfrak{D}(x^{\mu} - z^{\mu})d\tau, \tag{92}$$

where

$$\mathfrak{D}(x) \equiv \frac{\delta^4(x)}{\sqrt{-g(x)}}. \tag{93}$$

In case of no curvature in special relativity, the term $\sqrt{-g}^{\,-1}$ is not necessary, but it will be needed in case of general relativity in order to assure covariance of the delta function, thus we include it. Next, since the multipole moments are defined on the 3-dimensional spatial surface spanned by $\{e_i\}$, we set them on the general coordinates as follows,

$$\begin{aligned} M^{\rho_1\cdots\rho_n} &= e_{i_1}^{\rho_1}\cdots e_{i_n}^{\rho_n} M^{i_1\cdots i_n}, \\ J^{\mu\nu\cdot\rho_1\cdots\rho_n} &= e_a^{\mu}e_b^{\nu}\cdot e_{i_1}^{\rho_1}\cdots e_{i_n}^{\rho_n} J^{ab\cdot i_1\cdots i_n}. \end{aligned} \tag{94}$$

The third point is the Levi-Civita symbol ε^{ijk}. Because the spatial part of the comoving frame under consideration is spanned by $\{e_i\}$, the orthonormal vector to this spatial surface is v, and not u. Therefore, 3-dimensional ε^{ijk} satisfies the following relation with 4-dimensional anti-symmetric tensor

$$\varepsilon^{ijk}e_i^{\ \mu}e_j^{\ \nu}e_k^{\ \alpha} = v_{\lambda}\frac{\varepsilon^{\lambda\mu\nu\alpha}}{\sqrt{-g}}. \tag{95}$$

By using this relation, the relation between two types of the spin multipole moments $J^{\mu\nu\cdot\rho_1\cdots\rho_n}$ and $S^{\beta\rho_1\cdots\rho_n}$ is given by

$$J^{\mu\nu\cdot\rho_1\cdots\rho_n} = v_{\lambda}\frac{\varepsilon^{\lambda\mu\nu\alpha}}{\sqrt{-g}}g_{\alpha\beta}S^{\beta\rho_1\cdots\rho_n}. \tag{96}$$

Consequently, the energy momentum tensor becomes as follows by the coordinate transformation (91),

$$
\begin{aligned}
T^{\mu\nu} = & \sum_{n=0}^{\infty} \int d\tau\, \partial_{\rho_1\cdots\rho_n} (u^{\mu} u^{\nu} M^{\rho_1\cdots\rho_n} \mathfrak{D}) - \sum_{n=1}^{\infty} \int d\tau \partial_{\rho_1\cdots\rho_{n-1}} (u^{\mu} \dot{M}^{\nu\rho_1\cdots\rho_{n-1}} \mathfrak{D}) \\
& - \sum_{n=1}^{\infty} \int d\tau \partial_{\rho_1\cdots\rho_{n-1}} (u^{\nu} \dot{M}^{\mu\rho_1\cdots\rho_{n-1}} \mathfrak{D}) + \sum_{n=1}^{\infty} \int d\tau \partial_{\sigma\rho_1\cdots\rho_{n-1}} (u^{\mu} J^{\nu\sigma\cdot\rho_1\cdots\rho_{n-1}} \mathfrak{D}) \\
& + \sum_{n=1}^{\infty} \int d\tau \partial_{\sigma\rho_1\cdots\rho_{n-1}} (u^{\nu} J^{\mu\sigma\cdot\rho_1\cdots\rho_{n-1}} \mathfrak{D}) + \sum_{n=2}^{\infty} \int d\tau \partial_{\rho_1\cdots\rho_{n-2}} (\ddot{M}^{\mu\nu\rho_1\cdots\rho_{n-2}} \mathfrak{D}) \\
& + \sum_{n=2}^{\infty} \int d\tau \partial_{\sigma\rho_1\cdots\rho_{n-2}} (\dot{J}^{\sigma\mu\cdot\nu\rho_1\cdots\rho_{n-2}} \mathfrak{D}) + \sum_{n=2}^{\infty} \int d\tau \partial_{\sigma\rho_1\cdots\rho_{n-2}} (\dot{J}^{\sigma\nu\cdot\mu\rho_1\cdots\rho_{n-2}} \mathfrak{D}).
\end{aligned}
\tag{97}
$$

Finally in this section, we refer to conditions that the multipole moments must satisfy. Since the multipole moments are defined by (94) in the general coordinates, and 3-dimensional ε^{ijk} and 4-dimensional $\varepsilon^{\lambda\mu\nu\alpha}$ relate each other by the equation (95), the multipole moments must satisfy the following relation with the vector ν normal to the spatial part in the oblique frame:

$$
\begin{aligned}
& \nu_{\mu} M^{\mu\cdots} = 0, \\
& \nu_{\rho} J^{\mu\nu\cdot\rho\cdots} = 0, \\
& (\text{or } \nu_{\mu} S^{\mu\cdots} = 0).
\end{aligned}
\tag{98}
$$

Needless to say, $J^{\mu\nu\cdot\rho\cdots}$ also satisfy

$$
\nu_{\mu} J^{\mu\nu\cdot\rho\cdots} = 0, \tag{99}
$$

because of (95). These conditions are generalization of the well-known spin supplementary condition for the spin multipole moment to the higher multipole moments.

What we have seen above is the formulation of multipole particles in special relativity. Note also that the multipole particles are the sources of the gravitational field but they do not accept any action from the field at this stage. Namely, we have not introduced any interaction between the particles and the field yet. The introduction of the interaction will be done in the next section.

MULTIPOLE PARTICLE IN GENERAL RELATIVITY

Now, we will change the stage of the energy momentum tensor, formulated in section 6, from special relativity to general relativity. The shift from special relativity to general relativity can be easily achieved. We often apply the so-called "comma-to-semicolon" rule to various cases: for example, equations for scalar fields, electromagnetic fields and so on. This "colon-to-semicolon" rule means the change of ordinal partial differential to covariant differential. This rule indicates that all physics must be covariant under any general coordinate transformation. This rule is sometimes referred as *equivalence principle* conventionally though it is not entirely precise usage of the term of equivalence principle. By applying this rule to the energy momentum tensor for the multipole particles, we can naturally introduce the interaction between the particles and the gravitational field. The same process is known in the gauge theories of interactions. In such gauge theories, by changing a global symmetry to a local symmetry, all interactions are introduced by gauge fields. In other words, we introduce the interaction between the multipole particles and the gravitational field by "gauge principle".

By applying the "colon-to-semicolon" rule to the equation (97), we obtain the following energy momentum tensor,

$$\begin{aligned} T^{\mu\nu} = & \sum_{n=0}^{\infty} \int d\tau \nabla_{\rho_1 \cdots \rho_n} (u^\mu u^\nu M^{\rho_1 \cdots \rho_n} \mathfrak{D}) - \sum_{n=1}^{\infty} \int d\tau \nabla_{\rho_1 \cdots \rho_{n-1}} (u^\mu \dot{M}^{\nu\rho_1 \cdots \rho_{n-1}} \mathfrak{D}) \\ & - \sum_{n=1}^{\infty} \int d\tau \nabla_{\rho_1 \cdots \rho_{n-1}} (u^\nu \dot{M}^{\mu\rho_1 \cdots \rho_{n-1}} \mathfrak{D}) + \sum_{n=1}^{\infty} \int d\tau \nabla_{\sigma\rho_1 \cdots \rho_{n-1}} (u^\mu J^{\nu\sigma \cdot \rho_1 \cdots \rho_{n-1}} \mathfrak{D}) \\ & + \sum_{n=1}^{\infty} \int d\tau \nabla_{\sigma\rho_1 \cdots \rho_{n-1}} (u^\nu J^{\mu\sigma \cdot \rho_1 \cdots \rho_{n-1}} \mathfrak{D}) + \sum_{n=2}^{\infty} \int d\tau \nabla_{\rho_1 \cdots \rho_{n-2}} (\ddot{M}^{\mu\nu\rho_1 \cdots \rho_{n-2}} \mathfrak{D}) \\ & + \sum_{n=2}^{\infty} \int d\tau \nabla_{\sigma\rho_1 \cdots \rho_{n-2}} (\dot{J}^{\sigma\mu \cdot \nu\rho_1 \cdots \rho_{n-2}} \mathfrak{D}) + \sum_{n=2}^{\infty} \int d\tau \nabla_{\sigma\rho_1 \cdots \rho_{n-2}} (\dot{J}^{\sigma\nu \cdot \mu\rho_1 \cdots \rho_{n-2}} \mathfrak{D}), \end{aligned} \tag{100}$$

where the dots over the multipole moments mean covariant differential by the proper time of the particles:

$$\cdot = \frac{D}{D\tau} \equiv u^\mu \nabla_\mu. \tag{101}$$

In order to derive equation of motion for the multipole particles, we can use Bianchi identity, which is an integrable condition of Einstein equation,

$$T^{\mu\nu}{}_{;\nu} = 0. \tag{102}$$

However, it is intricate to derive the equation of motion from mere calculations of the divergence of the energy momentum tensor, except mass monopole particle. Therefore, we need more simple and more systematic ways to derive equation of motion. This algorithm is assembled by using some theorems by Tulczyjew.

Tulczyjew gave the following three theorems in his paper [21]:

Lemma 2. For an arbitrary point x in the spacetime and for an arbitrary tensor $a^{\mu\cdots}$ defined on the world line of a particle, the following equation holds,

$$\int_{-\infty}^{+\infty} \nabla_\alpha \cdots \nabla_\beta \nabla_\rho \left[a^{\mu\nu\cdots}(\tau) u^\rho(\tau) \mathfrak{D}(x - z(\tau)) \right] d\tau$$
$$= \int_{-\infty}^{+\infty} \nabla_\alpha \cdots \nabla_\beta \left[\frac{D a^{\mu\nu\cdots}(\tau)}{D\tau} \mathfrak{D}(x - z(\tau)) \right] d\tau, \tag{103}$$

where u^μ is 4-velocity of the particle whose position is denoted by $z(\tau)$.

Theorem 3. (*Tulczyjew*) Let $n^{\mu_1\cdots\mu_k\nu_1\cdots\nu_m}(s)$ be an arbitrary tensor on a world line and $N^{\nu_1\cdots\nu_m}$ be a tensor which is constructed from $n^{\mu_1\cdots\mu_k\nu_1\cdots\nu_m}$ as follows

$$N^{\nu_1\cdots\nu_m} = \sum_{k=0}^{n} \int_{-\infty}^{+\infty} \nabla_{\mu_1} \cdots \nabla_{\mu_k} \left[n^{\mu_1\cdots\mu_k\nu_1\cdots\nu_m} \mathfrak{D}(x - z(\tau)) \right] d\tau. \tag{104}$$

Then $N^{\nu_1\cdots\nu_m}$ can be reduced to a canonical form

$$N^{\nu_1\cdots\nu_m} = \sum_{k=0}^{n} \int_{-\infty}^{+\infty} \nabla_{\mu_1} \cdots \nabla_{\mu_k} \left[v^{\mu_1\cdots\mu_k\nu_1\cdots\nu_m}(\tau) \mathfrak{D}(x - z(\tau)) \right] d\tau, \tag{105}$$

where $v^{\mu_1\cdots\mu_k\nu_1\cdots\nu_m} = v^{(\mu_1\cdots\mu_k)\nu_1\cdots\nu_m}$ and $u_{\mu_1} v^{\mu_1\cdots\mu_k\nu_1\cdots\nu_m} = 0$.

Note that it is not necessary for $v^{\mu_1\cdots\mu_k\nu_1\cdots\nu_m}$ to be orthogonal to u^μ by the contraction with the indices ν's of $v^{\mu_1\cdots\mu_k\nu_1\cdots\nu_m}$. Moreover, $v^{\mu_1\cdots\mu_k\nu_1\cdots\nu_m}$ is not generally equal to $n^{(\mu_1\cdots\mu_k)\nu_1\cdots\nu_m}$.

Theorem 4. (*Tulczyjew*) If $N^{\nu_1\cdots\nu_m}$ in theorem 3 satisfies

$$\int_D N^{\nu_1\cdots\nu_m} K_{\nu_1\cdots\nu_m} \sqrt{-g}\, d^4x = 0, \tag{106}$$

for an arbitrary $K_{\nu_1\cdots\nu_m}$ and an arbitrary domain D, then all the coefficients $v^{\mu_1\cdots\mu_k\nu_1\cdots\nu_m}$ in the canonical form of $N^{\nu_1\cdots\nu_m}$ must vanish.

Because it is not difficult to prove these theorems, we omit them in this chapter. If necessary, please consult the author's paper [5].

We can derive the equation of motion for the multipole particles by some straightforward calculations following the algorithm below:

1. Calculate $T^{\mu\nu}{}_{;\nu}$,
2. Use the lemma 2 and the commutation relation of covariant derivative to transform $T^{\mu\nu}{}_{;\nu}$ into the canonical form in theorem 3,
3. Using theorem 4, you will obtain the equation of motion from the coefficient of the canonical form.

We will show some concrete calculations using this algorithm in the next section. In the rest of this section, we discuss the relation between the force and the nonlinearity of the field.

In case of electromagnetism, even if we construct a point charge with arbitrary multipole moments, we can not obtain the force that the particle receives. In other words, the point charge can make arbitrary motions. This situation is exactly the same with the case of the linearized gravitational field in special relativity. However, the particle can not move arbitrarily in case of general relativity. This difference comes from the nonlinearity that general relativistic gravity has. When we consider only linear fields, there is a powerful property, *superposition principle*. Owing to this principle, we simply add every field created by individual particles and can obtain a field that is created by many particles. However, when the field has nonlinearity, we can not simply add plural fields created by particles. The field equation can not be satisfied if we adopt simple addition. Namely, the nonlinearity requires that the fields created by the individual particles must receive some deformation in order to connect with each other smoothly and to consistently satisfy the field equation. This means that the particles receive the reaction caused by the deformation of the fields. In other words, every object feels every force as a compensation of the nonlinearity of the fields.

The above argument is used in the study of back reaction by emitting gravitational waves [22]. For instance, we consider a small spherical black hole on a background gravitational field. The existence of the black hole causes perturbations on the background spacetime. On the black hole space time, perturbations from curved background arise. By adjusting the perturbations to connect these two spacetime smoothly, the motion of the black hole on the curved background spacetime will be determined. Such perturbative approach usually needs very complicated calculations. Our approach is more simple, but as a matter of fact, the basic concept is common to both approaches. In our approach, the usage of the delta function manages to paste the perturbation to each other stealthily. This is a great advantage of our method.

APPLICATIONS

In this section, we apply the multipole particle description in the last section to some problems.

Monopole Particle

First, we consider mass monopole particles for the simplest case. The monopole particle is defined by setting $M \neq 0$ of $l = 0$ and all the other $M^L = 0$ of $l \geq 1$. Of course, all spin multipole moments are set to zero. It is usually referred to as, merely, *point particle*. In this simplest case, we ascertain the derivation of the equation of motion. According to the definition of monopole particle, the energy momentum tensor (100) becomes

$$T^{\mu\nu}_{\text{monopole}} = \int d\tau M u^\mu u^\nu \mathfrak{D}. \tag{107}$$

The divergence of this energy momentum tensor is

$$T^{\mu\nu}_{\text{monopole};\nu} = \int d\tau \nabla_\nu \left(M u^\mu u^\nu \mathfrak{D} \right). \tag{108}$$

Applying the lemma 2 to the right hand side of this equation, we get

$$\int d\tau \nabla_\nu \left(M u^\mu u^\nu \mathfrak{D} \right) = \int d\tau \frac{D}{D\tau}(M u^\mu)\mathfrak{D}. \tag{109}$$

According to the theorem 3, the right side hand of (109) concludes

$$\frac{D}{D\tau}(M u^\mu) = 0. \tag{110}$$

Here, because u^μ is the 4-velocity of the particle, u^μ satisfies $u^\mu u_\mu = -1$ and we obtain from (110),

$$\frac{DM}{d\tau} = 0, \tag{111}$$

$$\frac{Du^\mu}{D\tau} = 0. \tag{112}$$

The first equation (111) means the constancy of M. This is a necessary condition for M to be interpreted as the mass of the particle. The second equation (112) is exactly the geodesic equation. It is well known that objects with only masses move along geodesics.

Spinning Particle

Now, we consider the other simple case that a particle has its mass and its angular momentum. In this case, the energy momentum tensor is

$$\begin{aligned} T^{\mu\nu} = \int d\tau \Big\{ & \left(M u^\mu u^\nu - \dot{M}^\mu u^\nu - \dot{M}^\nu u^\mu \right) \mathfrak{D} \\ & + \nabla_\rho \left[\left(M^\rho u^\mu u^\nu + u^\nu J^{\mu\rho} + u^\mu J^{\nu\rho} \right) \mathfrak{D} \right] \Big\}. \end{aligned} \tag{113}$$

Here, we do not fix the mass dipole moment. It will be fixed at the end of the derivation of the equation of motion. In case of electromagnetism, the dipole moment has a physical meaning since there are two kinds of charge, positive and negative. However, in the case of the gravitation, we can change the dipole moment at discretion since there does not exist negative mass. The mass dipole moments depend on the definition of the center-of-mass of the object.

The calculation to get the equation of motion is slightly long but straightforward. The following is the concrete process of the calculation. To begin with, we set

$$A^{\nu\mu} \equiv M u^\mu u^\nu - \dot{M}^\mu u^\nu - \dot{M}^\nu u^\mu, \tag{114}$$

$$B^{\rho\nu\mu} \equiv M^\rho u^\mu u^\nu + u^\nu J^{\mu\rho} + u^\mu J^{\nu\rho}, \tag{115}$$

for convenience. In order to apply the lemma 2 easily, we decompose these $A^{\nu\mu}$, $B^{\rho\nu\mu}$ into the component parallel to the 4-velocity u^{μ} and the components perpendicular to it as follows,

$$A^{\nu\mu} = a^{\nu\mu} - u^{\nu} a^{\mu}, \tag{116}$$

$$B^{\rho\nu\mu} = b^{\rho\nu\mu} - u^{\nu} b_1^{\rho\mu} - u^{\rho} b_2^{\nu\mu} + u^{\rho} u^{\nu} b^{\mu}, \tag{117}$$

where we define the following valuables,

$$a^{\nu\mu} \equiv P^{\nu}{}_{\lambda} A^{\lambda\mu}, \tag{118}$$

$$a^{\mu} \equiv u_{\nu} A^{\nu\mu}, \tag{119}$$

$$b^{\rho\nu\mu} \equiv P^{\rho}{}_{\sigma} P^{\nu}{}_{\lambda} B^{\sigma\lambda\mu}, \tag{120}$$

$$b_1^{\rho\mu} \equiv P^{\rho}{}_{\sigma} u_{\lambda} B^{\sigma\lambda\mu}, \tag{121}$$

$$b_2^{\nu\mu} \equiv u_{\sigma} P^{\nu}{}_{\lambda} B^{\sigma\lambda\mu}, \tag{122}$$

$$b^{\mu} \equiv u_{\sigma} u_{\lambda} B^{\sigma\lambda\mu}, \tag{123}$$

and the projection operator $P^{\mu}{}_{\nu}$ regarding u^{μ} is defined by,

$$P^{\mu}{}_{\nu} = \delta^{\mu}{}_{\nu} + u^{\mu} u_{\nu}. \tag{124}$$

Now, the divergence of the energy momentum tensor becomes

$$T^{\mu\sigma}{}_{;\sigma} = \int d\tau \left\{ \nabla_{\sigma} \left[A^{\sigma\mu} \mathfrak{D} \right] + \nabla_{\sigma} \nabla_{\rho} \left[B^{\rho\sigma\mu} \mathfrak{D} \right] \right\}. \tag{125}$$

Let us analyze $\nabla_{\sigma} \left[A^{\sigma\mu} D \right]$ in (125) first. By applying the lemma 2 with (116) we obtain

$$\int d\tau \nabla_{\sigma} \left[A^{\sigma\mu} \mathfrak{D} \right] = \int d\tau \left\{ \nabla_{\sigma} \left[a^{\sigma\mu} \mathfrak{D} \right] + \frac{Da^{\mu}}{D\tau} \mathfrak{D} \right\}. \tag{126}$$

Second, we calculate $\nabla_{\sigma} \nabla_{\rho} \left[B^{\rho\sigma\mu} \mathfrak{D} \right]$ in (125). By using (117) and the commutation relation of the covariant derivatives, we have

$$\int d\tau \nabla_\sigma \nabla_\rho \left[B^{\rho\sigma\mu} \mathfrak{D} \right] = \int d\tau \left\{ \nabla_\sigma \nabla_\rho \left[b^{(\rho\sigma)\mu} \mathfrak{D} \right] + \frac{1}{2} [\nabla_\sigma, \nabla_\rho] \left(b^{[\rho\sigma]\mu} \mathfrak{D} \right) \right.$$
$$- \nabla_\rho \left(\frac{D}{D\tau} b_1^{\rho\mu} \mathfrak{D} \right) - [\nabla_\sigma, \nabla_\rho] \left(u^\sigma b_1^{\rho\mu} \mathfrak{D} \right) \tag{127}$$
$$\left. - \nabla_\sigma \left(\frac{D}{D\tau} b_2^{\sigma\mu} \mathfrak{D} \right) + \nabla_\sigma \left(\frac{D}{D\tau} (u^\sigma b^\mu) \mathfrak{D} \right) \right\}.$$

Since the parts operated by $\frac{D}{D\tau}$ in this result do not match the canonical form of the theorems 3 and 4, we must projectively decompose them regarding the direction of u^μ. Finally, we obtain

$$\int d\tau \nabla_\sigma \nabla_\rho \left[B^{\rho\sigma\mu} \mathfrak{D} \right] = \int d\tau \left\{ \nabla_\sigma \nabla_\rho \left[b^{(\rho\sigma)\mu} \mathfrak{D} \right] \right.$$
$$+ \nabla_\sigma \left[P^\sigma{}_\rho \frac{D}{D\tau} \left(u^\rho b^\mu - b_1^{\rho\mu} - b_2^{\rho\mu} \right) \mathfrak{D} \right]$$
$$+ \frac{1}{2} [\nabla_\sigma, \nabla_\rho] \left(b^{[\rho\sigma]\mu} \mathfrak{D} \right) - [\nabla_\sigma, \nabla_\rho] \left(u^\sigma b_1^{\rho\mu} \mathfrak{D} \right) \tag{128}$$
$$\left. + \frac{D}{D\tau} \left[u_\rho \frac{D}{D\tau} \left(u^\rho b^\mu - b_1^{\rho\mu} - b_2^{\rho\mu} \right) \mathfrak{D} \right] \right\}.$$

Getting the above results together, we have the divergence of the energy momentum tensor in the canonical form as follows,

$$T^{\mu\sigma}{}_{;\sigma} = \int d\tau \left\{ \nabla_\sigma \nabla_\rho \left[b^{(\rho\sigma)\mu} \mathfrak{D} \right] \right.$$
$$+ \nabla_\sigma \left[a^{\sigma\mu} \mathfrak{D} + P^\sigma{}_\rho \frac{D}{D\tau} \left(u^\rho b^\mu - b_1^{\rho\mu} - b_2^{\rho\mu} \right) \mathfrak{D} \right]$$
$$+ \frac{Da^\mu}{D\tau} \mathfrak{D} + \frac{1}{2} [\nabla_\sigma, \nabla_\rho] \left(b^{[\rho\sigma]\mu} \mathfrak{D} \right) - [\nabla_\sigma, \nabla_\rho] \left(u^\sigma b_1^{\rho\mu} \mathfrak{D} \right) \tag{129}$$
$$\left. + \frac{D}{D\tau} \left[u_\rho \frac{D}{D\tau} \left(u^\rho b^\mu - b_1^{\rho\mu} - b_2^{\rho\mu} \right) \right] \mathfrak{D} \right\}.$$

By applying the theorem 4 to this result, we obtain

$$b^{(\rho\sigma)\mu} = 0, \tag{130}$$

$$a^{\sigma\mu} + P^\sigma{}_\rho \frac{D}{D\tau} \left(u^\rho b^\mu - b_1^{\rho\mu} - b_2^{\rho\mu} \right) = 0, \tag{131}$$

$$\frac{D}{D\tau}\left[a^{\mu}+u_{\rho}\frac{D}{D\tau}\left(u^{\rho}b^{\mu}-b_{1}^{\rho\mu}-b_{2}^{\rho\mu}\right)\right]-[\nabla_{\sigma},\nabla_{\rho}]\left\{\frac{1}{2}\left(b^{[\rho\sigma]\mu}\right)-\left(u^{\sigma}b_{1}^{\rho\mu}\right)\right\}=0, \quad (132)$$

as the equation of motion.

In order to get concrete expressions, let us substitute (116) and (118)-(123) for these equations (130)-(132). First, (130) becomes a trivial equation $0=0$ that is always satisfied. Next, the projection of (131) regarding u^{μ} concludes as

$$\dot{j}^{\sigma\rho}u_{\rho}=0. \quad (133)$$

Furthermore, by symmetrizing and antisymmetrizing the perpendicular components of (131) to u^{μ}, we obtain

$$P^{\nu}{}_{\mu}P^{\rho}{}_{\sigma}\left(M^{(\sigma}\dot{u}^{\mu)}\right)=0, \quad (134)$$

$$P^{\nu}{}_{\mu}P^{\rho}{}_{\sigma}\frac{D}{D\tau}\left(-J^{\sigma\mu}+M^{[\sigma}u^{\mu]}\right)=0, \quad (135)$$

respectively. Finally, by using the relation

$$R^{\mu}{}_{\alpha\beta\gamma}+R^{\mu}{}_{\beta\gamma\alpha}+R^{\mu}{}_{\gamma\alpha\beta}=0,$$

we reach the following result

$$\frac{D}{D\tau}\left[Mu^{\mu}-\dot{M}^{\mu}-\left(u_{\sigma}M^{\sigma}\right)\dot{u}^{\mu}\right]+R^{\mu}{}_{\gamma\alpha\beta}u^{\gamma}\left[J^{\alpha\beta}-M^{\alpha}u^{\beta}\right]=0, \quad (136)$$

from (132). This is our equation of motion for a spinning particle.

Now, we have not fixed the freedom of mass dipole moment yet. It is nothing else but the definition of the center-of-mass of an object. By defining dipole moment adequately, we can simplify the equation of motion for a spinning particle. For example, let us redefine the mass and the spin dipole moment as follows,

$$m\equiv M+\left(u_{\sigma}\dot{M}^{\sigma}\right), \quad (137)$$

$$S^{\mu\nu}\equiv 2J^{\mu\nu}-M^{\mu}u^{\nu}+M^{\nu}u^{\mu}, \quad (138)$$

and choose M^{μ} to hold

$$P^{\mu}{}_{\sigma}\dot{M}^{\sigma}+\left(u_{\sigma}M^{\sigma}\right)u^{\mu}\equiv\dot{S}^{\mu\sigma}u_{\sigma}. \quad (139)$$

The existence of M^{μ} that satisfies (139) is assured by the existence theorem for solutions of differential equations. In these settings, the equation of motion (136) becomes the so-called Papapetrou equation [8],

$$\frac{D}{D\tau}\left(mu^{\mu}+u_{\sigma}\dot{S}^{\sigma\mu}\right)+\frac{1}{2}R^{\mu}{}_{\gamma\alpha\beta}u^{\gamma}S^{\alpha\beta}=0, \tag{140}$$

$$\dot{S}^{\mu\nu}+\dot{S}^{\mu\sigma}u_{\sigma}u^{\nu}-\dot{S}^{\nu\sigma}u_{\sigma}u^{\mu}=0, \tag{141}$$

and the energy momentum tensor becomes

$$\begin{aligned} T^{\mu\nu}_{\text{spinning}} = \int d\tau \Bigg\{ & \left(mu^{\mu}u^{\nu}-\frac{1}{2}\frac{DS^{\mu\rho}}{D\tau}u_{\rho}u^{\nu}-\frac{1}{2}\frac{DS^{\nu\rho}}{D\tau}u_{\rho}u^{\mu}\right)\mathfrak{D} \\ & -\frac{1}{2}\nabla_{\rho}\left[\left(S^{\rho\mu}u^{\nu}+S^{\rho\nu}u^{\mu}\right)\mathfrak{D}\right]\Bigg\}, \end{aligned} \tag{142}$$

which agrees with the energy momentum tensor given by the article by Tulczyjew [21].

Of course, we can choose the other definitions of mass dipole moment M^{μ}. The simplest choice is $M^{\mu}=0$ as well as the case of Newtonian gravity. In this case, the equation of motion (136) becomes simpler as follows,

$$M\frac{Du^{\mu}}{D\tau}+R^{\mu}{}_{\gamma\alpha\beta}u^{\gamma}J^{\alpha\beta}=0, \tag{143}$$

$$\frac{DJ^{\mu\nu}}{D\tau}=0, \tag{144}$$

$$\frac{DM}{D\tau}=0. \tag{145}$$

Note that this equation has been known as the pole-dipole approximation of the Papapetrou equation. However, according to our derivation, this is not an approximate one but exactly a relativistic equation of motion for a spinning particle The energy momentum tensor in this case is given by

$$T^{\mu\nu}=\int d\tau\left\{Mu^{\mu}u^{\nu}\mathfrak{D}+\nabla_{\rho}\left[u^{\mu}J^{\nu\rho}+u^{\nu}J^{\mu\rho}\right]\mathfrak{D}\right\}. \tag{146}$$

We will use this to post-Newtonian equation of motion for spinning particle.

Application to an Approximation Method

Here we show that the multipole particle description can be useful in the post-Newtonian theory. For this purpose, we derive the post-Newtonian equation of motion for spinning particles. The reason for considering post-Newtonian equation, though we know the fully relativistic equation, is that the relativistic equations are generally of no use for the analysis of many-particle systems. This is because we need to solve both of the equations of motion and of the field simultaneously, and it is too difficult except for numerical methods. Hence, we must use any approximation method in order to take account of relativistic effects for many-particle systems such as binary systems.

Here, we will derive the equation of motion that describes the orbital motion of a binary system. We consider only the spin-orbit coupling force. This result at first post-Newtonian order is already known, for example, in the article by Damour [23]. The equation of motion at a higher post-Newtonian order is first obtained by H. Tagoshi, A. Ohashi and B. J. Owen [18] by using different energy momentum tensor from the one examined in this chapter but essentially equivalent.

In the post-Newtonian calculation, we need to expand the gravitational field up to a necessary post-Newtonian order. It is convenient to define the following energy momentum density,

$$\sigma = T^{00} + T^{ii}, \tag{147}$$

$$\sigma_i = T^{0i}, \tag{148}$$

where $T^{ii} = T^{11} + T^{22} + T^{33}$. Let us introduce some retarded potential defined by

$$V = \Box_{\mathrm{R}}^{-1}\{-4\pi G\sigma\}, \tag{149}$$

$$V_i = \Box_{\mathrm{R}}^{-1}\{-4\pi G\sigma_i\}, \tag{150}$$

where

$$\Box_{\mathrm{R}}^{-1}\{-4\pi G f(\vec{x},t)\} \equiv \int \frac{d^3y}{|\vec{x}-\vec{y}|} f(\vec{y}, t-|\vec{x}-\vec{y}|). \tag{151}$$

By using these valuables, the metric is expanded into the following form

$$g_{00} = -1 + 2V, \tag{152}$$

$$g_{0i} = -4V_i, \tag{153}$$

$$g_{ij} = \delta_{ij}(1+2V), \tag{154}$$

up to needed post-Newtonian order in order to calculate leading spin-coupling terms.

Now, let us use the following energy momentum tensor

$$T^{\mu\nu} = \int d\tau \left\{ M u^\mu u^\nu \mathfrak{D} + \nabla_\rho \left[\left(u^\mu J^{\nu\rho} + u^\nu J^{\mu\rho} \right) \mathfrak{D} \right] \right\}. \tag{155}$$

By substituting this energy momentum tensor into the definitions of the densities, we get

$$\sigma = \int d\tau \left\{ M\mathfrak{D} + 4 v^k J^{kl} \partial_l \mathfrak{D} \right\}. \tag{156}$$

$$\sigma_i = \int d\tau \left\{ M v^i \mathfrak{D} + J^{il} \partial_l \mathfrak{D} \right\}, \tag{157}$$

up to a needed order. Here, the spin supplementary condition, $u_\rho J^{\rho\nu} = 0$, is imposed on the spin dipole moment $J^{\mu\nu}$. Consequently, the retarded potentials are given by

$$V(\vec{x}) = \frac{M}{|\vec{x} - \vec{z}(t)|} + 4 v^k J^{kl} \partial_l \frac{1}{|\vec{x} - \vec{z}(t)|}, \tag{158}$$

$$V_i(\vec{x}) = \frac{M v^i}{|\vec{x} - \vec{z}(t)|} + J^{il} \partial_l \frac{1}{|\vec{x} - \vec{z}(t)|}, \tag{159}$$

where $\vec{z}(t)$ is the position of the particle. This result is the expression for one particle, but we may sum them up in order to get the potentials in case of many particles since the definition of the retarded potentials is linear to particles. Therefore, we obtain

$$V(\vec{x}) = \sum_a \left\{ \frac{M_a}{|\vec{x} - \vec{z}_a(t)|} + 4 v_a^k J_a^{kl} \partial_l \frac{1}{|\vec{x} - \vec{z}_a(t)|} \right\}, \tag{160}$$

$$V_i(\vec{x}) = \sum_a \left\{ \frac{M_a v_a^i}{|\vec{x} - \vec{z}_a(t)|} + J_a^{il} \partial_l \frac{1}{|\vec{x} - \vec{z}_a(t)|} \right\}, \tag{161}$$

in case of many particles. In this expression, a identifies the individual particle.

Next, let us expand the equation of motion in term of the retarded potentials. The equation of motion derived from the energy momentum tensor (155) is

$$M \frac{Du^\mu}{D\tau} + R^\mu{}_{\gamma\alpha\beta} u^\gamma J^{\alpha\beta} = 0. \tag{162}$$

Since this form is slightly unsuitable, we deform the following form,

$$M\alpha^i + M\left(\Gamma^i{}_{\alpha\beta} - v^i \Gamma^0{}_{\alpha\beta} \right) v^\alpha v^\beta + \left(R^i{}_{\gamma\alpha\beta} - v^i R^0{}_{\gamma\alpha\beta} \right) v^\gamma \frac{J^{\alpha\beta}}{u^0} = 0, \tag{163}$$

where

$$v^0 = 1, v^i = \frac{u^i}{u^0}, \tag{164}$$

$$\alpha^i = \frac{dv^i}{dx^0}. \tag{165}$$

According to (163), at the lower post-Newtonian order, the effect of the spin of the particle itself on the motion comes from the terms of the connection terms and the effect of the spin of other particles comes from the terms of Riemann tensors. By counting the post-Newtonian order of the potentials carefully, we only consider the following terms

$$\begin{aligned} M\alpha^i + M\left(\Gamma^i{}_{00} + \Gamma^i{}_{0j}v^j\right) + \left(2R^i{}_{00j}J^{0j} + R^i{}_{0jk}J^{ik}\right) \\ = M\alpha^i + M\left(\left[-4V_{i,0} - V_{,i}\right] + \left[-2V_{i,j} + 2V_{j,i}\right]v^j\right) \\ + \left(2\dot{V}_{,j}J^{ij} + 2V_{,ij}v^l J^{lj} + \left[V_{k,ij} - V_{j,ik}\right]J^{jk}\right), \end{aligned} \tag{166}$$

in order to derive the spin-orbit coupling terms of the equation of motion at the leading order. By substituting the potentials into this equation, we will obtain the equation of motion for the many-body system of spinning particles. However, there are some self-energy terms, and we must regularize them. In the higher post-Newtonian calculation, we need a complicated regularization procedure. However, it is sufficient to neglect them in our case. Thus, we obtain the equation of motion for two spinning particle system as follows,

$$\alpha_1^i = \left(\frac{M_2}{r}\right)_{,i} + \left[\left(\frac{2J_1^{il}}{M_1} + \frac{4J_2^{il}}{M_2}\right)v_{12}^k\left(\frac{M_2}{r}\right)_{,kl} + \left(\frac{4J_1^{kl}}{M_1} + \frac{4J_2^{kl}}{M_2}\right)v_{12}^l\left(\frac{M_2}{r}\right)_{,ik}\right], \tag{167}$$

where r is the orbital separation between the particles and v_{12} means the relative velocity. This consists of the well-known Newtonian gravity and spin-orbit coupling terms at the leading order. However, our expression is different from a form, for example, in Damour's article [23]. This comes from the definitions of angular momentum. If we relate them by

$$J_a^{ij} = \frac{1}{2}S_a^{ij}, \tag{168}$$

it becomes clear that they are equivalent to each other.

As we saw above, when we study the post-Newtonian physics of objects with internal structures, we need both the gravitational field created by the objects and the relativistic equation of motion. Since the energy momentum tensor uses delta functions, our multipole particle description has the following two advantages,

- it simplifies the calculations of potentials,
- it is possible to derive the equation of motion.

From this fact, you may notice that the multipole particle approach is useful in analyzing relativistic many-particle systems that consist of objects with various internal structures.

SUMMARY

In this chapter, we developed the multipole particle method to describe the motion of extended objects, or celestial objects that have various internal structure. In relativity, to consider spatial extent means only to consider a non-local physics. Therefore, if you simply take account of the extent of matter distribution, you will just end up with an insufficient description of the physics. However, the multipole particles method always enables local description of the physics, because it considers the physical quantities at only one point, *particle*. This is an important reason why we consider multipole particles. Furthermore, as we have seen in the last section, the multipole particle method uses Dirac's delta functions and the delta functions make many calculations, such as the calculation of the gravitational potentials in the post-Newtonian approximation, very simple. More than anything else, this method provides not only equations of motion but also the energy momentum tensor of the particles. The study of many-particles systems, such as binary systems, requires a complete pair of the equation of motion and the energy momentum tensor. Despite the advantages of the multipole particle approach, some questions will arise. we will discuss them in the rest of this section.

First, we discuss the ambiguity of the relation between the distribution of matter and the multipole moments of the multipole particle. As we inferred, the distribution of matter is generally non-local. On the other hand, we need to exclude the quantity defined by such non-local distribution from the description of the motion. This, in turn, made the relation between them unclear. However, is there any opportunity that we can find the distribution of matter in the actual situation? Indeed, we can probe the structure of celestial objects, such as neutron stars etc., with various theories, but it is impossible to probe it directly. What we can do is only to presume it by using theories and through observations. We only can obtain the knowledge of the multipole moments of the objects through observations of motion. Then, should we describe physics only by quantities we can observe? This is where we stand. For instance, in the case of observations of gravitational waves, we can observe a wave form of gravitational waves modulated by the orbital motion of the binary system (to be precise, we choose a best-fitted template of gravitational wave forms and regard the template as the actual wave form). What we can obtain from the wave form is the multipole moments and their changes. Therefore, we consider that the multipole particle description is necessary and sufficient. In addition, we note that the relation between the distribution of matter and the multipole moments of the multipole particle becomes clearer in post-Newtonian approximation.

Next, we consider a question that any point particle can not be a physical entity. Surely, it is difficult to consider any point particle as a physical object in general relativity since the point particle must become a black hole or a naked singularity. This means that the point where the particle exists can not be a point that belongs to the spacetime manifold. Then, isn't it strange that such a particle indicates the location of an object? Hyper functions on curved manifold are mathematically well-defined objects. Therefore, the use of delta functions in the energy momentum tensors does not cause a problem. Then, how can we interpret them physically? We defined the multipole moments of the multipole particles by the gravitational field that is created by the object on the way of constructing the multipole particle framework. One of the essentials of the concept of the multipole particle is to describe the motion of objects regardless of the details of matter distribution. In other words, neglect the internal structures of objects if the objects create an identical external gravitational field. Therefore, we do not need to care whether the object is a black hole or a naked singularity. Namely, the role of delta functions is to provide boundary conditions on a boundary that separate the inside and the outside of an object. In this sense, the particle settles itself at a virtual point of the background spacetime.

Finally, we refer to one of the unsolved problems so far. That is the spin supplementary condition. Almost all articles treating spinning particles seem to impose artificial conditions as the spin supplementary condition. Although the spin supplementary condition can not be derived from equations of motion or the other physical constraints, it is indispensable for the equations to be closed. Of course, it is well known that the motion becomes ill-behaved, if we impose random conditions. Nevertheless, at least as far as the author knows, there is no particular rule we need to follow. As we referred to in section 5, the spin supplementary condition indicates on which surface of simultaneity we defined the multipole moments of the objects. We can choose a surface of simultaneity arbitrarily, since we can impose any spin supplementary conditions independently of the equation of motion. Despite this arbitrariness, the spin defined on such an arbitrary surface affects the motion of the object in exactly the same manner. This fact seems to be quite strange. In our formulation of the multipole particle, a condition on v^{μ}, which determines the spin supplementary condition, seems to be derived as follows. Let us differentiate the spin supplementary condition below by time along the motion of the particle,

$$v_{\rho} J^{\rho\sigma} = 0. \tag{169}$$

Then, we obtain

$$\dot{v}_{\rho} J^{\rho\sigma} + v_{\rho} \dot{J}^{\rho\sigma} = 0. \tag{170}$$

By multiplying this by u_{σ} and by using (133),

$$\dot{v}_{\rho} J^{\rho\sigma} u_{\sigma} = 0. \tag{171}$$

is derived. Consequently, the equations of motion becomes mathematically inconsistent unless the spin supplementary condition use v^{μ} that satisfies (171) at least. Thus, we need to probe whether the condition (171) gives a sensible constraint on the spin supplementary condition as a future question to be solved.

REFERENCES

[88] *The Penguin Dictionary of Physics*, third edition (Penguin Books, 2000)

[89] S. Chandrasekhar, *An introduction to the study of stellar structure* (Dover Publications, 1963)

[90] S. Chandrasekhar, *Ellipsoidal figures of equilibrium* (Dover Publications, 1969)

[91] R. A. Hulse and J. H. Taylor, "Discovery of a pulsar in a binary system", *Ap. J.* L51 (1975)

[92] Akira Ohashi, "Multipole particle in relativity", *Phys. Rev.* **D68**, 044009 (2003)

[93] Leopold Infeld, "Equations of Motion in General Relativity Theory and the Action Principle", *Rev. Mod. Phys.*, 398 (1957)

[94] V. Fock, "Three Lectures on Relativity Theory", *Rev. Mod. Phys.*, 325 (1957)

[95] A. Papapetrou, "Spinning test-particles in general relativity. I", *Proc. R. Soc. London* A., 248 (1951)

[96] W. G. Dixon, "A Covariant Multipole Formalism for Extended Test Bodies in General Relativity", *Il nuovo cimento*, 318 (1964)

[97] W. G. Dixon, "Dynamics of extended bodies in general relativity I. Momentum and angular momentum", *Proc. Roy. Soc. Lond.* A., 499 (1970)

[98] W. G. Dixon, "Extended Bodies in General Relativity: Their Description and Motion." in *Isolated Gravitating System in General Relativity*, edited By J. Ehlers (North-Holland, Amsterdam, 1979), p.156

[99] A. Einstein, L. Infeld and B. Hoffmann, "The Gravitational Equations and the Problem of Motion", *Ann. Math.,* 65 (1938)

[100] Tadayuki Ohta, Hiroshi Okamura, Toshiei Kimura and Kichiro Hiida, "Physically Acceptable Solution of Einstein's Equation for Many-Body System", *Prog. Theor. Phys.*, 492 (1973)

[101] Misao Sasaki and Hideyuki Tagoshi, "Analytic Black Hole Perturbation Approach to Gravitational Radiation", *Living Rev. Relativity* 6, (2003), 6. Online article]: cited on http://www.livingreviews.org/lrr-2003-6/

[102] Luc Blanchet, "Gravitational Radiation from Post-Newtonian Sources and Inspiralling Compact Binaries", *Living Rev. Relativity* 5, (2002), 3. Online article]: cited on 25 Dec 2000 http://www.livingreviews.org/lrr-2002-3

[103] T. Tanaka, Y. Mino, M. Sasaki, M. Shibata, "Gravitational waves from a spinning particle in circular orbits around a rotating black hole", *Phys. Rev.* **D54**, 3762 (1996)

[104] Benjamin J. Owen, Hideyuki Tagoshi and Akira Ohashi, "Nonprecessional spin-orbit effects on gravitational waves from inspiraling compact binaries to second post-Newtonian order", *Phys. Rev.* **D57**, 6168 (1998)

[105] Hideyuki Tagoshi, Akira Ohashi and Benjamin J. Owen, "Gravitational field and equations of motion of spinning compact binaries to 2.5 post-Newtonian order", *Phys. Rev.* **D63**, 044006 (2001)

[106] Kip S. Thorne, "Multipole expansions of gravitational radiation", *Rev. Mod. Phys.*, 299 (1980)

[107] L. D. Landau and E. M. Lifshitz, *The Classical Theory of Fields*, fourth revised english edition (Butterworth-Heinemann, 1987)

[108] W. Tulczyjew, "Motion of multipole particles in general relativity theory", *Acta Phys. Pol.*, 393 (1959)

[109] Yasushi Mino, Misao Sasaki and Takahiro Tanaka, "Gravitational radiation reaction to a particle motion", *Phys. Rev.* **D55**, 3457 (1997)

[110] Thibault Damour, "The problem of motion in Newtonian and Einstein gravity", in *300 Years of Gravitation*, edited by Stephen Hawking and Werner Israel (Cambridge University Press, Great Britain, 1987), 128

INDEX

A

B

C

D

E

Q

R

S

T

U

V

W